1295

W9-BFY-861

A Treasure Hunting Text

Ram Publications
Hal Dawson, Editor

Modern Treasure Hunting

The practical guidebook to today's metal detectors; "how-to" manual that explains the "why" of modern detector performance.

Treasure Recovery from Sand and Sea

Tells any treasure hunter how to reach the "blanket of wealth" beneath sands nearby and under the world's waters.

Modern Electronic Prospecting

Explains in layman's language how anyone can use a modern detector to find gold nuggets and veins of precious metal.

Buried Treasure of the United States

Complete field guide for finding treasure; includes state-by-state listing of sites where treasure can be found.

Modern Metal Detectors

Advanced handbook for field, home and classroom study to increase expertise and understanding of metal detectors.

Successful Coin Hunting

The world's most authoritative guide to finding valuable coins with all types of metal detectors.

Getting the Most from your Grand Master Hunter

Specific instructions for getting "plus" performance from Garrett's new computerized microprocessor-controlled detector.

Treasure Hunter's Manual #6

Quickly guides the inexperienced beginner through the mysteries of full time treasure hunting.

Treasure Hunter's Manual #7

The classic book on professional methods of research, recovery and disposition of treasure.

Weekend Prospecting

Describes how to enjoy holidays and vacations profitably by prospecting with metal detectors and gold pans.

Gold Panning is Easy

Excellent field guide shows the beginner how to find and pan gold as easily as a professional.

Treasure Hunting Pays Off

An introduction to all facets of treasure hunting . . . the equipment, the targets and the terminology.

The Secret of John Murrell's Vault

Fictional story of a search that really took place for a treasure still awaiting discovery. A True Treasure Tale.

Garrett Guide Series — Pocket-size field guides

Find an Ounce of Gold a Day
Metal Detectors Can Help You Find Wealth
Find Wealth on the Beach
Metal Detectors Can Help You Find Coins
Find Wealth in the Surf
Find More Wealth with the Right Metal Detector

MODERN TREASURE HUNTING

by Charles Garrett and Roy Lagal

ISBN 0-915920-61-1
Library of Congress Catalog Card No. 88-06 2346
Modern Treasure Hunting
©Copyright 1988
Charles L. Garrett & Roy Lagal

All rights reserved. No part of this book may be reproduced or transmitted in any form or by any means, electronic or mechanical, including photocopying, recording or by any information storage and retrieval system, except in the case of brief quotations used in reviews and articles. For information address all inquires to Ram Publishing Company. Printed in Korea. Distributed in the United States of America.

First Edition Printing. September 1988

88   89   90   9   8   7   6   5   5   4   3   2   1

For FREE listing of RAM treasure hunting books write
Ram Publishing Co. ● P.O. Box 38649 ● Dallas, TX 75238

Book and cover designs by Mel Climer
Back cover photograph by Hal Dawson

Contents

About the Authors

Note from the Authors

1. Introduction   1

2. What is Treasure Hunting?   11

3. Characteristics of Treasure Hunters   15

4. Coin Hunting   23

5. Cache (Money) Hunting   43

6. Searching Buildings and Cabins   59

7. Searching Beach and Surf   71

8. Relics and Battlefields   97

9. Antique Bottles   103

10. Electronic Prospecting   113

11. Rocks, Gems and Minerals   133

12. Finding Gold with Pan and Detector   137

13. Metal Detector History   141

14. Modern (Computerized) Detector Development   151

15. Practical Theory   157

16. Detector Configurations   173

17. Searchcoils   179

18. Clothing and Accessories   189

19. Points to Ponder   197

20. Conclusion   211

About the Authors

It would be trite—but true—to say that readers of Ram books need no introduction to the authors of *Modern Treasure Hunting*. These two men have truly enriched the literature of treasure hunting with stories of their exploits, along with methodical "how to" instruction that has introduced countless novices to the world of metal detecting in the proper manner and improved the skills of many an advanced hobbyist or professional treasure hunter.

In this volume, these two metal detector veterans seek to introduce hobbyists and treasure hunters to the new instruments designed to take us into the 21st century. Both writers bring to this project a lifetime of experience, vast knowledge of metal detecting and keen interest in all phases of treasure hunting.

Charles Garrett, a treasure hunter since the 1940's, founded one of the most successful metal detector companies in the industry more than 25 years ago. A graduate engineer, he introduced discipline to the manufacture of metal detectors and generally raised the standards of all detector hobbyists. He has personally used a metal detector to search for treasure on every continent except Antarctica, and he has also scanned under the lakes, seas and oceans of the world. Known as the **Grand Master Hunter,** he has written of his experiences in many books, some of which have become accepted as virtual texts for the hobby.

Roy Lagal, too, has written widely, especially in his chosen fields of electronic prospecting and cache hunting. With his very livelihood dependent upon his prowess with a metal detector and gold pan, Roy has achieved success as a full-time prospector in the United States, Mexico and Canada for some 40 years. He has worked with Charles Garrett for the past 20 years in the design of new Garrett detectors and has helped develop metal detecting from a haphazard pastime to a veritable science. Throughout the hobby he is known as the **Father of Electronic Prospecting.**

These two men who have successfully worked together for so many years have joined their talents still another time to present to treasure hunters of the world this guidebook to *Modern Treasure Hunting.*

Hal Dawson, Editor
Ram Books

Authors' Note

Commercial and hobby metal detectors have been improved constantly since the first models were offered to the public soon after World War II. Yet, progress has been most intensive over the past dozen or so years. All of us associated with treasure hunting welcomed these advancements and the opportunities they provided. Yet, recent developments have *completely revolutionized* treasure hunting as we have known it for so many years. Anyone who seeks to use a modern metal detector effectively must be responsive to new ideas – and, ready to modify habits of a lifetime. New skills are needed to enable the modern treasure hunter to develop those techniques necessary for proper utilization of the 21st century instruments now available. That's right . . . **the 21st century instruments!** The day of microprocessor-controlled and computer-designed instruments is here!

These new computerized detectors controlled by microprocessors – especially those with which we are familiar from Garrett Electronics – represent a quantum leap forward. The two of us speak as treasure hunters as well as a detector company president and a lifetime prospector and cache hunter. We accept the responsibility as two individuals helping guide the development of new detectors to meet the challenges of the future. It behooves us both to remain abreast – truly, to stay ahead – of all industry developments.

The new computerized instruments have proved absolutely amazing in performance. They have improved our abilities to locate and recover treasure to an extent greater than we had ever dreamed possible. And now, it is our goal with this book to share with treasure hunters of all ages and all levels of experience the results of our laboratory and field research with these revolutionary new detectors. We want you to experience the same benefits we have already enjoyed.

Although we would probably be considered "expert" treasure hunters, this book is not written solely for those with our advanced skills. Any beginner can benefit from it. At the same time, however, we believe that even the most veteran users of metal detectors need this manual—especially as it discusses the new computerized instruments.

Thus, this is a **how to** field guide, designed to hasten the development of a beginner into a seasoned, successful treasure hunter and electronic prospector—in the shortest possible time.

For the beginner, then, the guide seeks to offer instruction in selecting and using the most modern metal detection equipment available. It also provides a complete understanding of all phases of treasure hunting and presents the knowledge required to choose a particular phase of the hobby in which to seek excellence. As a novice, you can develop a solid foundation that lets you advance as far as you wish in that area of the hobby in which you choose to specialize. At the same time, you will gain knowledge and proficiency in recovering other types of treasure that will prove beneficial when such opportunities present themselves . . . as they surely will!

Serving yet another purpose, this straightforward, down-to-earth, "skills" manual explains to the veteran how he or she can vastly improve detecting ability simply by recognizing capabilities of the new equipment and incorporating them in individual techniques.

Veterans of the hobby will find confusion surrounding the various types of new metal detectors removed as you are guided, step by step, through all aspects of the use of modern instruments. You will find their various features explained clearly and learn how the new detectors can help you become more successful in hunting coins, caches, relics, precious metals or whatever treasures you seek.

Quick success for the beginner and vastly improved skills for the veteran, therefore, are the goals of this field guide. Simply stated, it is our desire to help you get maximum satisfaction from this wonderful hobby that both of us have enjoyed for so many decades.

Material in this guide has been drawn from other books that we have written. Among these are *Detector Owners Field Manual, VLF-TR Metal Detector Handbook, Modern Metal Detectors* and *Treasure Recovery from Sand and Sea.* Using techniques from these older manuals that have proved successful over the years and over the world; this new guidebook will help you understand the basics of metal detector operation while leading you into the new computer age.

May you receive as much pleasure from reading this new guidebook as we have enjoyed in writing it . . . and, may it enable you to have an even fuller treasure pouch when we . . . see you in the field!

Charles Garrett   **Roy Lagal**

Chapter 1

Introduction

This advanced guidebook on treasure hunting with the most modern detectors and searchcoils available will come quickly to the point in all of its presentations. The goal throughout will be to explain in clear and simple terms how modern metal detectors can enhance anyone's treasure hunting abilities.

This guidebook will consider *only* the newest and most modern equipment. Old style detectors, such as the beat frequency oscillator (BFO) and transmitter/receiver (TR) instruments, will *not* be discussed. They are simply gone . . . obsolete. They have been replaced . . . most adequately! Modern computers make possible today's metal detector circuitry that should be considered intricate only in its simplicity. Circuitry that controls modern detectors features microprocessors and other exotic components that perform feats we could only dream of just a few years ago.

Detectors can now think, and they possess memories to remember those vital aspects of treasure hunting that the hobbyist so often forgets.

This guidebook will begin with an explanation to broaden your understanding of the *basic* capabilities of "universal" metal detectors utilizing very low frequency (VLF) circuitry. Our example will feature the legendary Master Hunter ADS which has undergone field use for millions of hours by thousands of treasure hunters and prospectors. We will consider those techniques used by skilled metal detectorists to achieve maximum performance from such instruments. Then, you will be able to understand how designs of the new, computerized equipment were built upon a foundation of vast experience gained through

1

use of the earlier VLF equipment.

This book will not be highly technical. It simply presents the subject of treasure hunting with a metal detector in a manner that has proven successful for the authors over the years as they have instructed untold numbers of hobbyists . . . young and old, novice and veteran.

Before progressing farther, let us assure the reader that the new computerized equipment is so easy to use that even a beginner can achieve instant success with it. Truly, the new pre-programmed computer memory and microprocessor controls are, without a doubt, the finest circuitry ever designed for metal detecting equipment. What is amazing, however, is that these new instruments are the easiest of all detectors to operate. More about this later.

Yet, even with such simplicity, the ultimate capabilities of the new microprocessor-controlled equipment are so advanced that they challenge the most accomplished of metal detectorists to delve as deeply into the intricacies of the hobby as he or she could wish. In short, the capabilities of the new and modern metal detectors are truly limitless. In using them we are bounded only by our finite imagination and our desire to continue learning their field capabilities.

You will soon observe that this book clearly shuns using most of the many "alphabet-soup" terms that have been coined to describe various features of metal detectors. For the most part, these terms are outmoded and truly meaningless. Some, such as BFO and TR, are generic. Mostly, however, the terms – and the initials that described them so confusingly – were the product of manufacturers who tried to sell their equipment through the use of exotic-sounding terms that "proved" one particular detector was superior to another.

Before you ask the question, let me hastily answer: mea culpa. Yes, Garrett Electronics also participated in the mania of creating initials, but the following quiz represents our supreme effort to "kick the habit!"

A Metal Detector Quiz

To amuse veteran hobbyists and explain to newcomers the extent of the craze of using initials in the metal detector industry, join in the following examination. Below are 22 sets of letters that have been used to describe various features of metal detection equipment. Next to each, write your best understanding of what words the initials stand for and/or the feature of metal detecting to which they apply. Also, you might try to remember which company used the letters first.

Veteran hobbyists should be able to give answers to at least 15 of these terms. If you know less than eight, you're probably a newcomer or an amateur, still striving diligently to learn the "how-to" material presented in this book and spending many hours in the field sharpening your skills.

BFO	**RF (Two Box)**
TR	**PRG**
MF	**IB**
MPD	**GC**
SPD	**COAX**
ADS	**COP**
GEB	**CONC**
GB	**VLF**
PI	**PB**
SC	**CDC**
ESI	**P2L**

Almost unbelievable, isn't it? And, this list is by no means complete! Can you see how confusing the field of metal detecting had become? Sometimes, it was difficult even for those of us in the industry to understand the complete meaning of certain terminology, much less believe the boasts that so profusely accompany a new term.

Believe me, metal detecting does not have to be complicated, and it isn't . . . with the new computerized, microprocessor-controlled instruments!

Yes, the computerized metal detectors are here to stay. Their capabilities are too great for us not to welcome them to our hobby . . . indeed, to greet them with open arms. They permit us treasure hunters to overcome obstacles today that prevented us from finding certain treasures yesterday. Now is the time to learn about them and their capabilities.

In this guidebook you will find that all useless jargon has been stripped away. Using simple terms, we will present and explain the features, capabilities and "how-to" techniques of modern equipment. They will be explained when and as you need them as you progress in the various phases of treasure hunting. These features should be for YOUR benefit, not to sell a detector by making its operation seem complicated or exotic. If you cannot understand how to use a feature on any detector, that feature is worthless to you, no matter how "important" it may have been presented in published literature.

This is truly a book about 21st century treasure hunting and prospecting. Equipment that we will discuss can 1)search deeper into the earth; and 2)precisely analyze what it finds in a manner that was never before possible. The book is current – not years late–because it comes directly on the heels of the new computerized equipment. Some of you may already own such detectors. Many others of you probably plan to buy one. Perhaps not, you say? Believe me, you will. And, as one automobile slogan used to suggest, *Eventually . . . why not now?*

Yes, the computerized metal detectors are here to stay. Their capabilities are too great for us not to welcome them to our hobby . . . indeed, to greet them with open arms. They permit us treasure hunters to do things today that could only be dreamed of yesterday. Now is the time to learn about them and their capabilities. Now is the time to use and understand the new equipment. You can be successful like no others before you for the simple fact that these earlier treasure hunters were

Facing
Modern computerized detectors with microprocessor-controlled circuitry, modern as tomorrow's space shot, transport treasure hunters into the 21st century.

Over
Charles Garrett presents one of his Freedom detectors to the Countess of Strathmore at historic Glamis Castle in Scotland, where he searched for treasure.

using obsolete equipment.

Some or our knowledge and advice, especially as it pertains to the laws of physics and certain detector characteristics, may sound familiar. This is to be accepted. The laws of physics and mathematics do not change. Neither does good detector usage.

Modern Treasure Hunting seeks to build a bridge connecting the old and reliable techniques with today's new, advanced detectors and their increased capabilities. Those professionals among you readers will mentally acknowledge his or her learned understanding of old methods and quickly grasp the new. You who are still novices have the opportunity to "get in on the ground floor" with the modern equipment by learning techniques that have already proven successful in the past. These will lead to a good understanding of the easy-to-master advanced capabilities of today's computerized equipment.

Now, walk with us across the bridge that this book seeks to build. Don't leave your old detector experience behind . . . or your knowledge of it. But, be prepared to learn new techniques . . . and to expand your horizons. We urge you to be open minded and willing. New doors will be unlocked for you, and you will gain knowledge guaranteed to make treasure hunting with a metal detector more exciting . . . more profitable . . . more fun than ever before!

Facing
The authors field test detectors under widely varying conditions at locations literally all over the world such as this site in Egypt.

Over   Top
Bob Lilly, former football great with the Dallas Cowboys, discusses prospecting techniques with Charles Garrett at his office in the Garland plant.

        Bottom
Mel Fisher, left, describes to Charles Garrett the pleasure he experienced when his crew found the *Atocha*, a 16th century Spanish galleon loaded with treasures from Mexico. Looking on is Fay Feild, electronic designer who worked with Mel for many years to perfect the design and implementation of special underwater metal detection equipment.

The Answers

BFO	*Beat Frequency Oscillator*
RF (Two Box)	*Radio Frequency*
TR	*Transmitter/Receiver*
PRG	*Pulse Readout Gradiometer*
MF	*Mineral Free*
IB	*Induction Balance*
MPD	*Magnetic Phase Discrimination*
GC	*Ground Control*
SPD	*Synchronous Phase Discrimination*
COAX	*Coaxial*
ADS	*Automatic Dectection System*
	Advanced Detection System
COP	*Co-Planar*
GEB	*Ground Exclusion Balance*
CONC	*Concentric*
GB	*Ground Balance*
VLF	*Very Low Frequency*
PI	*Pulse Induction*
PB	*Push Button*
SC	*Searchcoil*
CDC	*Computer Designed Circuitry*
ESI	*Electro-Static Induction*
P2L	*Phase Locked Loop*

Chapter 2

What is Treasure Hunting?

Before we go into more detail about how to use modern metal detectors to find treasure, let's take an overall look at the hobby of treasure hunting today. Let's try to find out just what it's all about and just what constitutes a treasure hunter.

Now, the purpose of *Modern Treasure Hunting* is not to tell "treasure stories." Various other Ram books have plenty of fascinating true treasure tales you should read about. And, if you insist on using older, non-computerized detector models, there are other Ram books that describe this equipment and tell you how to get the most from it.

We certainly aren't suggesting that you abandon that old favorite metal detector with which you have found so much treasure. No, indeed! Use it in good health and enjoy! But, remember that you are missing countless discoveries because of your inadequate equipment. That's right, as difficult as it may be for you to believe . . . with your old detector you are scanning *right over* treasures deeper and probably more valuable than those you are recovering.

To study treasure hunters let's divide them into three categories: amateur, semi-professional and professional.

Most of you will probably fall into the amateur category . . . at least, at first. And, many of you will never want to leave the amateur category . . . which is good! Amateurs generally have more fun with metal detectors than the more experienced operators who are trying to "prove" something or earn a living. An amateur hunts treasure with a metal detector for the sheer enjoyment of doing it. Any objects of value that may be found are but a bonus.

Believe us, however, when we predict that newcomers will uncover more treasure than you can imagine. Too many treasure discoveries have occurred in this hobby for us to doubt the ability of any one of you to become proficient to the point of being successful.

The amateur treasure hunter decides what treasures he or she wants to find and then goes after them. To be successful in attaining any treasures, however, the novice must make certain of possessing:

– Adequate *knowledge* and
– The right *equipment*.

You must gain the knowledge, first, before you can truly know which equipment will be needed. Books such as this are an important first step for the treasure hunter in learning about the hobby and the correct equipment with which to pursue it. Let us urge, however, that you try to study books that deal with *modern* equipment and techniques and not learn about detectors that are already obsolete. Remember that the modern detectors are many times more capable than those BFO's, TR's and other old instruments whose praises are being sung so lovingly and delightfully.

Where else can you uncover knowledge about modern detectors? Your metal detector dealer can be an important source of such information. Treasure hunting clubs will include members with all levels of experience, from novice on up. Most of them are always interested in new equipment. You can learn much from these other treasure hunters, plus enjoy the fellowship of our great fraternity. The treasure hunting magazines contain interesting stories and will often discuss new detectors. Their ads will prove especially helpful in introducing you to the new detectors.

Once you know what sort of treasure you want to look for and how you want to go after it, you'll be able to select the **right** equipment. Please let us emphasize the importance of using the right kind of detector wherever and whatever you are hunting. A new book in the *Garrett Guides* series is entitled *How to Find More Treasure with the Right Detector.* We wish that every treasure hunter could read this little booklet. Newcomers to the hobby would learn so much, and old-timers could test the strenth of their own opinions. This book tells in the simplest possible terms what kind of metal detector to buy for the various kinds of hunting. We urge all metal detector hobbyists to read it.

When seeking that *right* detector for your type of hunting, one of your major considerations must be value. Of course, value is the relationship between what an item costs and how well it performs. High quality detectors are expensive, but you'll never "lose money" in buying them because you'll be using equipment that is the most capable of finding the treasure you seek. On the other hand, if you purchase a cheap, off-brand model, you may not realize a return on your money and time. Inexpensive detectors just cannot perform as accurately and efficiently as equipment that is properly designed and field tested adequately.

We wish that every hobbyist could hunt with the new Grand Master Hunter. This instrument with computerized, microprocessor circuitry is the smartest, most capable metal detector that has ever been offered to the treasure hunting public. The world's first "thinking" metal detector, it is the most sophisticated instrument ever designed. Yet, most of its capabilities and functions are automatic, and it operates with the touch of a single control pad.

If you prefer a detector from another manufacturer, we urge you to get a new model with modern features. Sure, it may cost you a few dollars more than their older models, but you'll gain this money back many times over with the increased treasure you'll find.

Semi-Pros and Professionals

We listed three kinds of treasure hunters, but we've spent a great deal of time on the amateur. That's as it should be since less experienced treasure hunters comprise a primary audience for this book. At the same time, we want to introduce the new and modern instruments to the more advanced hobbyists and professional treasure hunters who have not yet tried them. Or . . . who don't believe the claims made for these new detectors!

Just who are the semi-professional treasure hunters? For the most part they are men and women who make their living in a field closely allied to treasure hunting. Two good examples are certain metal detector dealers themselves and the operators of diving shops. These semi-pros sell eqiuipment to treasure hunters, and they spend their spare time using it themselves. You'd expect those people selling equipment to keep up with the latest developments. For the most part, they do, and they will certainly make this equipment available to their customers. How effectively the semi-pro adapts his or her techniques to the use of the new instruments with their increased capabilities will determine the true level of professionalism that each treasure hunter achieves.

At the top of the profession are those very few who make their living hunting treasure. These are men (and some women) of whom seldom is heard. They are individuals who work quietly and alone, for the most part. They may pursue a treasure literally for years without even picking up a detector. Research is their keyword. They go looking for a specific treasure only when they are relatively certain where it is. Their finds may be few, but they may also be large.

The professional treasure hunter needs no urging to try out new equipment and new ideas. They understand that someone seeking a large, missing treasure – in competition with others just as experienced and proficient – needs every advantage possible. That new, deeper-seeking, more sensitive detector may be the instrument that can find the treasure that other detectors have been passing over.

Although we have stated that the amateur or beginning treasure hunter presents the primary audience for this new book, we believe that it will be welcomed as well by the semi-professional and professional treasure hunters who seek to learn all they can about new detectors and more creative ways of using them. Presenting such information is a goal of *Modern Treasure Hunting*.

Chapter 3

Characteristics of Treasure Hunters

This chapter discusses the basic attributes of a success-ful treasure hunter – those quirks of his or her nature that differentiate the treasure hunter from anyone else. Now, it's true that just about anyone who likes the outdoors and possesses even a little curiosity will enjoy "scanning around" with a metal detector. Some healthy exercise is guaranteed, and financial rewards may be more tangible, depending upon the dil-igence (and luck) of the hobbyist.

Becoming a successful treasure hunter is far more difficult. Expertise with a metal detector is never enough; it takes a "spe-cial" kind of person. The working title of this chapter was *"Can You Become A Treasure Hunter?"* We sought to explain the psy-chological make-up of a successful treasure hunter; then, to ask you to answer truly if you could meet these requirements. That sounded so difficult that we changed our title to *"SHOULD You Become A Treasure Hunter?"*

Finally, we decided that the prerequisites of treasure hunting were neither that difficult nor limiting. We present 12 of them as **Characteristics of a Treasure Hunter.**

1. Desire to find lost treasure.

This characteristic is basic to any good treasure hunter. It demands both the *belief* that lost treasure is waiting to be found and the *curiosity* or *greed* to want to find it. It is this **desire** that continually brings a coin hunter back to parks and playgrounds time and time again, even while he or she is seeking out new places to hunt. The belief that treasure is waiting to be found draws relic hunters miles off the map into wilderness areas or keeps them searching known battle areas for "that" rusty Civil War pistol. It is the firm conviction that nuggets await that draws the electronic prospector and his metal detector into the gold fields. Curiosity "wonders" how the nuggets will look when made into jewelry, while greed "measures" their value in the marketplace.

Common sense dictates that lost coins and jewelry abound on any beach where people play or have played. It is the DESIRE to find them that translates this commonsensical knowledge into a deep belief that jewelry, coins and the like are simply awaiting discovery by your metal detector. The successful treasure hunter continually says to him or herself that if he or she does not find the treasure, it will be found by somebody else. This characteristic, therefore, is a *desire that transcends belief!*

2. Love of historical knowledge.

This important characteristic often spells the difference between average, hard working treasure hunters and those whose brilliant successes make them outstanding. Historical knowledge leads to grand recoveries, but historical knowledge is not achieved without hard work . . . work with facts and dates that can sometimes seem dull and uninteresting.

The successful treasure hunter, however, thrills to research that might uncover a cache lost for long years or even a recent hidden lode of treasure. They enjoy reading historical books and histories of events at all levels, local as well as national. They seek out Indian villages, settlers' campgrounds, ghost towns and scenes of battles. They stop to read the historical markers on the highways, the plaques on obscure buildings. They enjoy gossip from the past about "whatever happened to soandso" and "where his money went" and such. They welcome stories about loot missing from old crimes. Always, they are seeking the natural disaster, the lost battlefield, the quiet, wealthy miser.

Cache hunters especially display this characteristic. They can begin their long journey with but a single fact. Sometimes it is as obvious as, "A wealthy man farmed here and died without any heirs." More often, it is simply, "There was a farmhouse here once." From this single fact – or rumor! – they seek data until such a full scenario is developed that they need only follow it with their metal detector to discover where to dig.

3. Patience.

This simple virtue presents such obvious rewards that it is envied by all. How many have prayed to their gods for patience? Then, demanded that it be granted to them "right now!" Patience is so important and vital that it is needed by all – and needed immediately.

It is a trait that must be included in the make-up of any successful treasure hunter, for all will come to *know* that many disappointments generally precede "striking it rich." Certainly, failure is not *expected*; indeed, each spadeful of earth is lifted away with the expectation that treasure awaits. The successful treasure hunter, however, works to transform the fruitless efforts of today into building blocks or stepping stones that will lead to tomorrow's worthwhile goals.

Whenever a novice turns on a metal detector, he or she should always remember that it takes *time* for an individual to become proficient in *any* endeavor.

The successful treasure hunter certainly enjoys the fruits of today's recoveries, but understands that much effort is required to plant seeds of the trees that will eventually bear this fruit.

4. Interest in analysis.

This trait, too, is related to curiosity because the successful treasure hunter must develop the ability to be inquisitive about places and things and to enjoy analyzing data about them. False leads must be discarded even as true leads are pursued. All facts about a potential recovery must be studied and analyzed. Let's say that you have been "assured" that a treasure was buried atop a mountain. How did it get there? Why did someone climb so high to bury it? Why wasn't it hidden at the base of the mountain? Maybe it was? **Find out!**

Discovering treasure need not be hard. The successful treasure hunter who is keenly interested in analysis knows that treasures are simply buried and await discovery. The best way to locate them is to seek direct answers to simple questions. An understanding of the importance of analysis helps develop these questions.

5. Lack of fear of hard work.

This trait must naturally sound somewhat negative since nobody really "likes" hard work. The successful treasure hunter, however, truly has no fear of it. And, that's what it takes! Money doesn't fall from trees or wash up on the beach. Many hours and days of scanning with a metal detector is usually required before a truly worthwhile prize can be discovered. Before you stoop to recover "that" target, you will have stooped countless other times and dug what seems like pound after pound of trash.

Success rarely comes easy. For the successful treasure hunter, it comes accompanied by what seems like gallons of perspiration pouring off the brow and pound after pound of soil that must be washed from clothes. The successful treasure hunter has the patience to understand that for success to be worthwhile, it must be worth working for.

6. Readiness for travel.

Success demands that you be always ready to follow the results of your curiosity, your research, your analysis and your hard work . . . even it it takes you miles away across county and state lines. Part of the thrill of research and analysis is to uncover clues that point to treasures far away.

The successful treasure hunter always displays faith in such efforts by following the clues they unearth . . . and following them without procrastination. That's important . . . being *ready!* How many times has a treasure hunter said, "Well, the next time I go to this-or-that-place, I'll look into that good lead." Then, years pass by, and that "next time" eventually finds that the "good lead" turns out to be a deep, freshly dug hole or an empty washtub in an old well.

7. Interest in the work of others.

The successful treasure hunter is genuinely interested in any treasure story that is encountered – no matter how it turns out. You must be prepared to spend time studying the successes and failures of others. Personal experience is absolutely necessary, but you should also benefit from the lifetimes of experiences of others that is available through books and videotapes.

The successful treasure hunter has a good reference library and uses it regularly to learn more about all phases of the hobby. Even when you disagree with someone, reading the ideas and opinions of that person challenges your imagination, and you profit from having been called on to question your own beliefs.

Professionals have willingly and gladly shared their lifetimes of experience with you. They have been honest because they have no reason to lie. Benefit from their experiences. Perhaps you can use one of their failures as a stepping stone toward your success.

8. Interest in seeking companions.

Your first choice should be the local metal detector dealer. Try to benefit from his knowledge and experience both in metal detector techniques and research efforts. Listen closely; you may be surprised what you learn. You're interested in the message, not the messenger.

Don't be afraid to ask for help. The person you ask may have once sought the same answer.

Look for the opportunity to join a treasure hunting club that will let you pool your knowledge. Certainly, you're not going to give out your good leads, nor can you expect to receive them from others. You can expect to meet people with similar interests, people from whom you can learn.

Don't be despondent if you can't find a metal detector club in your area. Organize one! Don't know how? Talk to your metal detector dealer, or write to Garrett.

9. Specializing in one type of hunting.

The field of treasure hunting is a broad one, indeed. And, almost all successful treasure hunters are truly generalists (See #12). They started out, however, by selecting a specific type of treasure to seek . . . then they pursued that type of treasure with absolute determination and diligence.

Do you seek coins? Are you going to cache hunt? Will you be seeking relics or gold? Do you plan to hunt on the beach or in ghost towns?

Choose your field of interest; then, study it carefully. Read stories and instruction manuals about that type of hunting. Try to talk to those who have been successful hunting the treasure you desire. Select sites with that specific target in mind. Study equipment designed to seek out such treasure.

10. Interest in treasure hunting equipment.

The successful treasure hunter is keenly interested in all types of metal detectors, their capabilities and their possible limitations. One manufacturer may be favored, but the products of all are studied regularly through personal use and by reading field test reports in the trade press and by reading the advertisements.

A good rule of thumb for any treasure hunter, novice or professional, in choosing a detector to buy is, first, to select the best equipment available for the type hunting you have chosen and based on what you can afford. Then, purchase a more expensive model of the same equipment. This does not mean that the successful treasure hunter always uses the most expensive equipment available. He or she does use, however, a true, all purpose or universal type detector designed for hunting multiple types of targets.

Certainly, if a specific type of target is chosen such as coins, beach hunting or electronic prospecting, specialized detectors for those targets are available.

Remember, too, that the successful treasure hunter is interested in the laboratory and field testing (See below.) that a piece of equipment has undergone . . . not just in its advertising claims.

11. Love for field testing.

The successful treasure hunter can't wait to get any new or different detector out in the field. What do they *claim* it can do? Now, what can I *make* it do? These are the questions that are asked. The answers to these questions supported by extensive field testing enable all treasure hunters, novice as well as professional, to understand the limits of their capabilities. The goal, then, becomes to extend these capabilities, to "squeeze" more from the detector than its manufacturer intended. Only extensive testing can make this possible.

Before taking any detector into the field, the successful treasure hunter first studies all literature supplied with it by the manufacturer. Incidentally, the following *Owner's Manual Test* is one that should be applied to any detector:

What does the Owner's Manual tell you about a detector? Does it tell you how to use it? Or, does it contain some meaningless statistics and diagrams? You don't buy a detector so that you can repair it; you buy it to search for treasure! Is the Owner's Manual printed in a format designed to go into the field with you or is it made for a desk drawer? In other words, is the Own-

20

er's Manual designed to be used or to be filed?

Before you buy a detector, ask to see the Owner's Manual that accompanies it. You may be surprised.

12. Desire to hunt all types of treasure.

This trait supplements #9, above, rather than conflicts with it. In theory, after you have become successful in one specific field you will naturally want to expand your expertise into other areas of treasure hunting.

What is more likely, however, is that such an event will occur when you least expect it. You will somehow be exposed to a treasure, a treasure story or a treasure site while you're still learning about your initially chosen targets. Yet, you understand how procrastination can be fatal to any treasure hunter's chances for success. If you've studied the literature of treasure hunting properly, you'll know how to go after your new target. And, that's our advice: Go!

Some may call it luck, but success will come to those who are prepared to recognize opportunities and who have trained themselves to follow through to achieve results.

During the study of your chosen field, make up your mind you are going to become the very best. But, always welcome the opportunity to branch out and study other fields of interest. Even though you have decided to become the very best coin hunter, do not hesitate to follow up on leads to a good cache. Know the size searchcoil you will need and be familiar with the thinking of someone who buries a cache.

Don't be afraid of the beach or the gold fields. Just make certain you have the right kind of equipment and know the general principles for using it. Basic theories of metal detector usage apply to all fields. It is in the area of specialization that you will learn little tricks from others and develop those tricks you learn on your own.

There you have **12 Characteristics of a Treasure Hunter.** Recognize yourself? Because most of these traits are born into an individual, you may possess some of them you've never before recognized because you've never needed them. Give yourself a chance. Within you may possibly lie the *makings of a successful treasure hunter!*

Chapter 4
Coin Hunting

Much of the material in this chapter will be all-encompassing. It will pertain to virtually every phase of treasure hunting. This is appropriate because coin hunting is the most popular recreational use for a metal detector. This statement can be made with no fear of contradiction. In fact, just about every instrument ever manufactured or sold will be used to hunt for coins at one time or another. Because all detector manufacturers understand this, all detectors are designed to find coins. It must be admitted honestly that many, many treasure hunters scarcely look for anything more than coins, no matter what kind of detector they are using.

Searching for coins can be so simple that even small children can participate. All that is needed for success is a quality metal detector and just a little research to locate good coin hunting spots. Yet, finding rare and older coins in difficult locations can be a complex task, indeed; so complex that the finest and most modern detectors in the hands of experienced operators are required.

The subject of *where* to find coins is one on which entire books have been written. And, this matter of "where" is vitally important to the treasure hunter. The simple fact is that *you cannot find coins where they do not exist!* Consider, if you will, the angler who selects a likely looking spot and patiently fishes there with no success. After a reasonable length of time, he will pack up his tackle and move on to another spot. On the other hand, we have known detector hobbyists who dig up trash target after trash target and complain that their detector won't locate coins. Because they are detecting small pieces of junk, the detector certainly would find coins which are larger. These detectors are not *missing* any coins; there simply are none there.

When this happens to the experienced coin hunter, he or she quickly moves to another spot, tests it and continues with this procedure until an area is discovered that produces coins to locate and recover. Always remember that no matter how "super-sensitive" or "deep-seeking" your detector might be, it cannot discover what does not exist. This rule applies equally to the areas of beach hunting, prospecting, cache hunting or any other phase of treasure hunting.

How a Detector Finds Coins

There's no magic to the way a metal detector can locate a coin in the grass, buried deeply in soil or rocks or in the surf. It's all a matter of electronics. A metal detector is simply an electronic device that detects the presence of metal, primarily through the transmission and reception of radio wave signals.

A metal detector is not an instrument (geiger counter) that detects energy emissions from radioactive materials. It is not an instrument (magnetometor) that measures the intensity of magnetic fields. It does not "point" toward coins or any other kind of metal; it does not measure the abundance of metal. A metal detector simply detects its presence and reports this fact.

Facing
Another prize from a modern, computerized detector! This "thinking" instrument is so simple and easy to operate; just press a single touchpad!

Over
Before coin-hunting Freedom detectors were introduced, Charles Garrett tested them in England. He is shown discovering an 18th century penny there.

When a coin is made of metal – and, most of them usually are – a metal detector will signal the location of this coin when it is beneath its searchcoil. How all this comes about is a somewhat more complicated story.

Knowing how a gasoline engine operates really doesn't make you a better driver. Similarly, it isn't necessary to understand all the scientific principles of metal detection for you to be able to use a detector to find coins. But, everyone knows that understanding an internal combustion engine and a little about how it makes the wheels turn helps make you a better operator of a motor vehicle. In the same manner, understanding the how and why of metal detection may make you a better coin hunter. Since *better* coin hunters find *more* coins, we will seek to explain in laymen's language just how a metal detector locates coins as well as all other metal.

Like any other metal object, a coin is detected essentially by the transmission and reception of radio wave signals. This is the scientific "principle" that governs operation of all metal detectors. If they're all the same, you may quickly ask, why do some cost so much more. One good question is followed by another. What distinguishes quality metal detectors from those that don't seem to find many coins or much of anything?

Facing   Top
This handful of coins was a real prize for one treasure hunter. Note the digging tool and headphones that he used to insure top coin hunting performance.

Bottom
Coins must be found where people have lost them. Here, a Freedom detector is used to search a field often crowded with people watching athletes perform.

Over
Financial rewards for a treasure hunter come in many shapes and sizes. Modern, computerized detectors are finding prizes that older instruments overlooked.

Primarily, it's the methods by which the detectors transmit signals and the sophistication with which the signals are received and interpreted. In a word, it's largely in circuitry.

When a radio signal is produced in the searchcoil of a metal detector, an electromagnetic field is generated that flows out into the surrounding medium, whether it be earth, rock, water, wood, air or any other material. This electromagnetic field doesn't particulary seek coins or any other kind of metal; it just flows out into the air, earth, water or whatever medium is present. But, the "lines" of this electromagnetic field penetrate metal whenever it comes within the pattern of the detection path. The extent of this pattern depends to a large degree upon the power used to transmit the signal and the resistance of the medium into which the signal is transmitted.

The electromagnetic field generated by transmission from a detector's searchcoil causes something called "eddy currents" to flow on coins detected by this field. Generating these currents on the coin causes a loss of power in the electromagnetic field and this power loss can be sensed by the detector's circuitry. Electromagnetic field lines passing through a coin and generating eddy currents on it, thus, distort or change the characteristics of the normal electromagnetic field. This distortion is the second of the clues that alert a detector to the nearby presence of metal.

These currents and their resulting distortion of the electromagnetic field are, as we said, sensed by a metal detector. At the same time, eddy currents on the coin generate a secondary electromagnetic field into the surrounding medium, which gives the detector still another clue. A receiver in the searchcoil detects these signals at the same time the loss of generating power is being detected. Circuitry of the metal detector simultaneously interprets all these sensations and generates appropriate signals to the operator. The detection device instantly reports that a coin (or, some sort of metal) appears to be present beneath its searchcoil.

Electromagnetic signals cause eddy currents to flow on the surface of any metal object (or mineral) having the ability to conduct electricity. Precious metals such as silver, copper and gold have higher conductivities and, appropriately, more flow of eddy currents than iron, foil, tin or other less desirable minerals. Since metal detectors can "measure" the amount of power that is used to generate eddy currents, the detector can "tell" which metals are the better conductors.

Quite simply, the quality of signals generated, received and interpreted by the metal detector and the ability of the coin hunter to act upon them determines the difference between "digging junk" and finding coins.

Unfortunately, the difference between junk and coins is rarely that simple.

Penetration of the electromagnetic field into the "search matrix" (that area over which a metal detector scans) is described as "coupling." Such coupling can be "perfect" into air, fresh water, wood, glass and certain non-mineralized earth.

Since life is seldom perfect, the search matrix "illuminated" by a metal detector (through transmission and reception of signals) contains many elements and minerals – some detectable and some not, some desirable and some not. A metal detector's electronic response at any given instant is caused by *all* conductive metals and minerals and ferrous non-conductive minerals illuminated in the search matrix by the electromagnetic field. Detection of minerals is, in most cases, undesirable.

And, wouldn't you know it? Two of the most undesirable are also two of the most common: natural iron (ferrous minerals) found in most of the earth's soil and wetted salt found in much of the earth's water. Not only do these minerals produce detection signals, but they inhibit the ability of instruments to detect metal.

When iron minerals are present or near the search matrix, the electromagnetic field is upset and signals are distorted. Iron mineral detection, therefore, presents a major problem to manufacturers and users of metal detectors. Although detection of such minerals may be desirable when a prospector is seeking black sand or ferrous magnetite that could contain gold or silver; it is a nuisance to the hobbyist who is looking for coins.

A primary design criterion of any detector, therefore, must be to filter or eliminate responses from undesirable elements, informing the treasure hunter only of those from desirable objects. This is accomplished in a variety of ways depending

31

upon the type of metal detector.

Ground Balance and Discrimination

Two terms, *ground balance and discrimination,* are used to describe the characteristics of a modern detector that enable it to seek out only desirable targets while ignoring trash and junk.

Electronic engineers achieve these characteristics through various methods of circuitry which properly manage the normal electrical phase relationship among resistive, inductive and conductive voltage. Phase shifting is a phonomenon basic to the understanding of electricity. Management of it to enable a specific metal detector to "dial out" iron mineralization or other undesirable targets, while still permitting the discovery of coins, involves highly proprietary knowledge and circuitry protected by U.S. patents. One of the authors and other Garrett engineers, incidentally, hold many of these patents, including a number that are primary in the manufacture of metal detectors.

Searchcoils

The most practical searchcoil for coin hunting is one approximately eight inches in diameter. This will probably be the "general purpose" size searchcoil with which your detector came equipped. We recommend that you use this size searchcoil for scanning most parks, playgrounds and other coin hunting areas. It is the best general purpose, all-around searchcoil. Many of you probably wonder about using an even larger coil for hunting coins. Our recommendation is that under certain conditions, yes, you can.

Some of you are probably asking just when should you use the larger searchcoil. Good question! After you scan an area with a modern detector using the standard-size coil, you may find you are detecting deep targets that give you only a faint signal, even with headphones. If so, you're detecting at the outer limits of that all purpose searchcoil, and you may be missing coins that lie deeper. You need more detection depth, and it's available with a searchcoil 12 inches or so in diameter. After scanning such an area thoroughly with the 8-inch searchcoil, go over it again with the larger coil, scanning very slowly and using headphones. If there are deeper coins to be found, you'll detect them!

There are also smaller coils to consider! At Garrett our smaller coil approximately four inches in diameter is called a "Super Sniper." You will find it particularly effective for coin hunting in trashy areas, in "tight" locations or around metal fences and playground equipment.

Because this coil is smaller, it inspects less area at any given time and, therefore, analyzes fewer potential targets. It is vital for you never to forget that your detector will merely report to you what is beneath its searchcoil at a specific point in time. If several targets are present, it will attempt to report that fact. If the area you are searching is filled with trash, targets are likely to be predominantly undesirable. When several of these trash tragets are beneath a searchcoil, signals may be confusing and you could overlook a coin. Because the smaller searchcoil covers less area, chances for "detection" of multiple targets are less likely.

Since the "Super Sniper" is smaller, it suffers less from nearby metal interference and can be used closer to metal fences, buildings and equipment than the larger coils.

One of the greatest coin-hunting tips that many treasure hunters put to good use is to *search with this smaller Super Sniper searchcoil*. Many areas that you will want to search contain a tremendous amount of junk . . . small bits and pieces of trash of all kinds, both ferrous and non-ferrous. Remember that your detector's circuitry will be continually analyzing all targets beneath its searchcoil at all times. When there are several targets at detectable depths, they are read simultaneously. Thus, you may occasionally hear numerous little blips (rejected target sounds) along with, perhaps, even an occasional single coin tone. To work these locations proficiently, we encourage you to use the Super Sniper.

Even though its narrow diameter lets you scan a much smaller area at any specific instant, it offers excellent detection depth. When the items detected in a smaller area of ground are considered by your detector's circuitry, it can do a much better job of analyzing individual targets, letting you pick out the good coins amongst the junk trash. Try the small searchcoil! You'll be convinced after only a short time of its great value. These small "Super Sniper" searchcoils are the most popular accessory coils sold by manufacturers.

It is important that whatever searchcoil you use be Faraday-shielded to make certain that interference from weeds and grass does not distort or blank out faint signals from deeply buried coins. You should also be aware of the environmental protection offered by your searchcoil before you use it in wet areas or submerge it.

All Garrett Crossfire searchcoils, for instance, are fully submersible to the connector and are Farraday-shielded. More information on this important subject of searchcoils can be found in Chapter 17.

Headphones

For maximum success a coin hunter should use headphones whenever searching with a metal detector. They are essential in noisy areas, such as the beach or near traffic. They enhance audio perception by bringing the sound directly into your ears while masking outside interference.

Just about anybody will hear weaker sounds and detect deeper targets when quality headphones are used. At Garrett we offer several selections in varying weight sizes and configurations, our most popular being stereo types that cover both ears. Adjustable volume headphones permit the operator a wide variance of audio adjustment without degrading the sound quality.

While speaking of sound and headphones, here's a tip about audio. Reducing sound volume to silent on a detector is accompanied by loss of detection depth and sensitivity. You may have read differently, but this is our opinion based on years of experience and success in the fields and waters:

Always operate your detector with the audio control adjusted so that just a faint sound comes from the speaker.

Then, when your detector signals a target, you will know it . . . and you'll always know that the detector is operating. Headphones allow this "threshold" to be set even lower, giving improved performance. You should experiment with your detector's threshold to determine the best audio level for you.

Modern Detectors

The first hobbyist ever to use a metal detector probably searched for coins . . . at least, occasionally. Every detector since manufactured has been used in a similar fashion. Success achieved with these earlier detectors varied widely, depending not only on the instrument itself but the proficiency of the hobbyist.

In the continuing development of metal detectors, improvement of two features – ground balance and discrimination – has contributed most significantly to the increased success possible today with modern detectors. Now that we have modern instruments it is difficult to realize that we were actually capable of finding coins with the older detectors, many of which lacked both of these features entirely.

It would be impossible to state which of the two features is more important. It is impossible to hunt in highly mineralized soil without circuitry that permits precise ground balancing. At the same time the deepest seeking detector is almost useless in trash or littered areas without adequate discrimination circuitry.

Before beginning any search, ground balance your instrument as precisely as possible to existing conditions. Many of the modern detectors feature automatic ground balancing, and peculiarities of this feature are discussed later in this volume. Some other detectors offer a choice between automatic and manual ground balancing. We recommend that when you are given such a choice that you first try the automatic ground balancing and give the detector's circuitry a chance. You might be surprised at its accuracy!

By all means, carefully follow manufacturer's instructions. If you should ever have trouble ground balancing your instrument, do not hesitate to consult with the dealer who sold you the detector or with its manufacturer. Remember to check ground balance from time to time while you are scanning, especially if you change hunting locations.

Discrimination

Here's another subject about which much has been and continues to be written. Modern deepseeking detectors are excellent for locating coins at great depths. Then can also detect tiny ferrous and non-ferrous objects so small the eye can hardly see them. Because they can detect nails at fantastic depths, they are in constant demand by loggers, sawmill operators and surveying crews. When properly ground balanced over highly mineralized soil, they can detect targets to surprising depths.

The level of discrimination to be "dialed in" is a continuing subject of interest. **Dig all targets** is the advice given by many treasure hunters. Yet, in a trashy area you could spend all day recovering trash (often very deeply) to find just a few coins. Precise discrimination circuitry available in modern detectors solves this problem for you.

Here's a tip on setting discrimination controls that might help you find more coins. Turn your discrimination to the *lowest setting possible*. On dials this is usually counterclockwise . . . as far to the left as they will go. Scan any area you have chosen for a few minutes or until you have dug a couple of dozen targets. Analyze these targets . . . study them carefully! Determine exactly what trash you are recovering. Then, set your discrimination controls absolutely no higher than necessary to reject the most troublesome of the little pests.

For instance, if you have located a dozen bottlecaps but only two pulltabs, set your detector to reject only bottlecaps. Do not set the controls to reject pulltabs; be content to dig a few of them for the greatest success. One axiom of treasure hunting you must learn: to have the greatest success, you must *dig some trash!*

A good rule to remember is to use as little discrimination as possible. Every time you increase the amount of trash you are eliminating by increasing your discrimination level you take the risk of losing good targets – coins, in this case.

No Discrimination

We previously offered advice given by some treasure hunters about "digging all targets." Well, it's an accepted fact that some good targets are "masked" when discrimination is used. Professional treasure hunters often search in the All Metal or similarly named modes to achieve the most depth possible and to locate the greatest number of coins, even though considerable trash is also dug.

Here's one tip that professional treasure hunters deploy to the fullest when their detector has a meter that correctly identifies all detected targets no matter what mode of operation is selected. Many of the modern detectors, especially the computerized models, feature such a meter. With it you can, therefore, scan in the All Metal mode and detect with audio all targets, both junk and treasure, at maximum depth. Once you have detected an object, simply glance at the meter to determine if you wish to dig or not. If the pointer falls in the "out-of-range" section of its scale, by all means dig the target. It could be a valuable object that is simply too deep to be read properly by the discrimination circuitry.

When scanning in the All Metal mode, there are other methods you can employ to increase your coin-hunting success. After you get a signal and have the target pinpointed, return to the mode which permits discrimination. If the audio signal increases in loudness or does not change, dig the target. It will probably be worthwhile. If the signal decreases or you hear a junk blip, don't bother to dig.

Any conductive trash that is between your searchcoil and a coin will affect the searchcoil's magnetic field and the effect of the coin's eddy currents on that field. From just listening to the target you may decide it is junk. Not so; obviously, in every case! Practice by distributing small junk targets of conductive metal at shallow depths with a coin placed more deeply. You will quickly notice that the interference from the smaller, but closer, junk targets reduces or distorts the signal from the coin. Here we go again! Sometimes, it may pay big dividends to go ahead and dig all targets to make certain you reach deeper and more valuable coins.

But, continue to pay particular attention whenever *any* audio signal is heard. Scan carefully around the spot . . . back and forth, side to side . . . listening carefully for that "coin" signal. Sometimes, coins when partially "hidden" by junk will produce a "coin" tone only at one tiny spot.

Searching where rare and valuable coins have been discovered can be very rewarding. You'll need to go more slowly for most profitable results, and you'll probably once again want to search in your All Metal mode or with very little discrimination.

Many operators will adjust their discrimination controls differently when seeking older coins or when searching in a supposedly worked-out area. Rejection should be adjusted to exclude only nails, because you do not want to miss any of the tiny, marginal coins with their heavy patina or low conductivity metal alloy content. Other operators adjust for bottlecap rejection, seeking to miss none of the small rings or nickels.

Let us repeat: The *less* discrimination you select, the *greater* chance you will have to get deep coins missed in previous searches.

Trashy Areas

We've discussed the use of smaller, Super-Sniper searchcoils to hunt in areas containing excessive amounts of trash. Now, let's talk about those areas that have just been considered too trashy for any searching. Select such a trashy area . . . a busy picnic ground or roadside park. Adjust discrimination control to reject pulltabs, and adjust tuning to achieve slightly more audio than usual. Operate your detector with the searchcoil approximately two to four inches above the ground. If the abundance of targets continues to be a problem, decrease sensitivity. Your goal is to move the searchcoil in slow sweeps and maintain a fairly steady audible response.

Listen carefully! Bad targets will cause the audio to decrease, and the good ones will cause perhaps only a slight increase because of your reduction in sensitivity. After just a little practice, you will find this method very effective on shallow coins. You may probably miss some small rings and nickels because of your discrimination setting, but this method opens up to you areas that are so trashy they may never have been searched before . . . certainly not searched satisfactorily.

Some operators seek better response and depth on coins by adjusting their discrimination to eliminate only nails. This will prevent your audio from decreasing as much, especially when trash targets with higher conductivity are detected. At this setting much of the trash you detect will be shallow. Thus, quick retrieval and identification of these shallow bottlecaps and pull-tabs will be easy. This method prevents most coins from being "masked" by junk targets.

As usual, experimentation and practice will guide you. We encourage you to compare various control settings and to carefully analyze target responses at different search sites. All things considered, no two sites are the same; consequently, a few minutes (or hours) spent analyzing any given area will probably result in greater success for you.

Automatic Circuits

Detectors with certain automatic circuits have a rebound or overshoot phenomenon of which you should be aware. Audio response, either postive or negative, will be followed immediately by a signal of opposite polarity. In other words, if the detector is discriminating and the searchcoil is passed over a junk target, the audio will decrease. As the searchcoil glides past this junk, the audio will rise sharply like a good positive signal. It is, however, only the "overshoot" effect produced when automatic tuning is used. Similarly, when you encounter a good target, the audio will increase over the target before falling back to become momentarily silent.

Practice until you become familiar with this characteristic. Bury a bottlecap and a coin about an inch deep and about a foot apart. With discrimination set to reject the bottlecap move the searchcoil over both targets at normal scanning speed and note the responses. You should be able to distinguish the difference easily.

Continue your practice sessions over known ground until you feel capable. Then, move to an area where extremely trashy conditions prevail. This and the other methods just presented are used successfully by many experienced operators in high trash areas to produce good results. It is simply a matter of learning to recognize and sort out "false" overshoot signals, target masking and other detector phenomena. Scanning at a constant speed—even slower than usual—and *taking advantage of the many tried and proven tricks* will permit you to achieve better accuracy and success.

Digging Tools

Many different types of tools are being used every day to recover single coins speedily. Some diggers and probes are designed especially for treasure hunters. Others are adaptations of the familiar icepick, screwdriver, garden trowel, bayonet or hunting knive. Each treasure hunter seems to have favorites of her or her own. Of course, the tool that you use must be effective, and it must not damage grass and soil conditions beyond that condition where they can be immediately repaired.

A slender, sharp-pointed object such as a screwdriver and a slim, thin probe are probably the most useful for recovering coins, especially in areas where care must be taken with turf. Careful probing helps you locate the coin precisely. Then, press the screwdriver, slightly to the side of your coin, into the ground a little deeper than the coin. Careful upward motions with the screwdriver under the coin will cause it to work to the surface. A bend in the end of the screwdriver converts the tool into an efficient "scoop." This retrieval method works best in sandy soil. Make certain that both the end of the probe and the screwdriver are blunt and will not scratch or otherwise damange coins.

A hunting knife is one of the more common tools carried by many treasure hunters. It can be used for "plugging" grass either to recover coins or before digging deeper holes. Plugging dry grass may cause it to die. When plugging, cut the plug larger than your coin target (about three-inch sides) and cut on only three sides. Then, lift and fold the plug outward and retrieve your coin. Replace the plug in the hole and press down firmly with your foot. Large grass plugs cut only on three sides and tamped down are less likely to be pulled out by lawnmowers. Considerably more information about digging tools can be found in Chapter 18.

Two other points are important when considering digging tools and recovery:

– Always scan over your holes. Just because you've recovered a good coin from a hole doesn't mean that there can't be one or more coins still in it.

– Protect your hands. Always wear gloves . . . at least, on the hand you use to dig into the hole. You might encounter a razor blade, broken glass or other sharp-pointed object.

Conclusion

As stated at the beginning of this chapter, much of its material pertains to all phases of treasure hunting. Metal detectors operate the same no matter what target is being sought. Your searchcoil will send and receive its signals regardless of what is beneath it. Our suggestions concerning the importance of ground balance and discrimination techniques apply universally . . . whether you're seeking coins in a park, gold nuggets in the desert or high country or buried treasure in the surf.

Still, it is coin hunting that draws the attention of most hobbyists, and the discovery of coins is always welcome, no matter what other targets are being sought. We urge you to study this chapter carefully . . . almost as carefully as you study the Owner's Manual for your detector. Follow its guidelines in an area where coins are likely to be present, and you will be successful!

That is why the hobby of coin hunting is so popular. Anyone can be successful . . . at just about any time and in practically any place!

Sportsmen are adding a quality metal detector to their complement of outdoor equipment. The avid fisherman does not have to become a treasure hunter just to search for coins. He brings a detector on his fishing trips and uses it to fill in hours of leisure when "they ain't biting" or when his arms grow tired from pulling in so many "big 'uns."

Similarly, anyone can carry along a metal detector to add this interesting and profitable hobby to their primary outdoor sport. This is true of the hunter, snowmobiler, back-packer, horseback rider and golfer. If you're new to the hobby of metal detecting, we ask that you think back to all the places you have followed another outdoor interest. Honestly consider the potential for discovery of coins and other buried treasure in these areas, and you will admit that you probably left behind a lot of treasure that a metal detector could have helped you find.

How about members of your family? Could they spend time enjoyably scanning for coins while you pursue another activity?

Let this chapter on coin hunting open the door to a magnificent and interesting hobby for you and those with whom you enjoy the great outdoors. Take advantage of this opportunity to join the great worldwide fraternity of treasure hunters!

Chapter 5

Cache (Money) Hunting

C ache hunting is *different*. Always remember that . . . no matter how successful you've been at finding coins . . . no matter how much jewelry you've dug out of the surf.

Money caches are truly the big prizes of treasure hunting, but many metal detector hobbyists never realize the vast differences between hunting for them and all other types of treasure. Thus, they are continually unsuccessful in their efforts to "hit the jackpot." They never seem to be able to recover caches of money, weapons or other valuable objects. And, they just can't seem to understand why.

We remember two nationally known treasure hunters who developed quite a reputation for their expertise with metal detectors. Through advertising and publicity they obtained a number of good treasure leads and information relating to cache sites. They generated considerable publicity about their proposed searches for caches and other "really big" prizes. They spent a great deal of their own and other people's money to pursue some of the leads, and they were successful in recovering a few coins and shallow relics. At first, their limited recoveries received the same high-powered publicity as their search plans. Eventually, the publicity and their fame died out.

As far as we can determine, the pair never recovered a substantial cache. Why? One reason is obvious: they were using small, coin-hunting searchcoils. Also, they were probably using the same techniques and procedures that had proven successful when they were recovering small items at relatively shallow depths.

Many experienced cache hunters would have paid good money for some of the information and leads these two squandered. How many valuable caches did this pair scan right over without knowing it? What a waste!

Different Type of Hunting

Cache hunting is different! Following are simple basic rules that have proven successful in this exciting phase of treasure hunting:

– Use a cache hunting detector and the *largest searchcoil* available.

– Conduct *extensive research;* you can never know too much about your target and the individual(s) who hid it.

–Be *patient* throughout your effort, from planning to scanning to recovering . . . and even after you dig up your prize.

–Never forget that while your target is generally big, it may be *deep* and, thus, more difficult to locate.

Certainly, we do not suggest you forget or ignore the tech-

Facing
When searching for a cache, a large coil on a modern computerized detector can often prove effective to find prizes that have been overlooked earlier.

Over
This valuable box of old coins is typical of the prizes that can be found by a cache hunter who carries out research and is patient in his recovery efforts.

niques you have already developed with your metal detector. As stated earlier, the laws of physics do not change. Rules for ground balancing and audio tuning that were true when you were hunting coins in the park will prove equally valid when you're seeking a cache in a deserted ghost town.

Let's consider some of the factors that are always important. All of these enter into successful recovery of deeply buried caches:

– Geographic location of the treasure site;
– Ground condition of the site and vegetation covering it;
– Mineral content of the soil;
– Physical size of the cache (generally overestimated!);
– Depth of the cache;
– Changes that might have occurred at the site since the cache was buried;
– Your detector and its searchcoil.

Misjudgment of any of the above factors may prevent sucessful recovery of the prize you seek. Of all these "stumbling blocks" we are convinced that the one mentioned last – your detector and its searchcoil – prevents more treasure hunters from discovering caches than any other.

You may be asking, "How can you criticize my reliable detec-

Facing   Top
Ruins of this old stone building are carefully searched with a metal detector to discover any items of conductive metal (gold) still concealed in walls.

Bottom
Author uses a Master Hunter detector to scan the walls of a pyramid in Egypt as part of his continual testing program for Garrett detectors.

Over
Depth Multiplier attachment which converts a Master Hunter into a **super** deepseeking detector also eliminates detecting small targets and trash.

tor, 'Old Betsy?' It's found all those coins for me. It located Roman relics when I took it to England. Why, it even found a nugget in the desert!"

Even so, there is a good chance that your reliable detector may be *almost worthless* for locating a deep cache – especially if you use a general purchase searchcoil. To recover large, deep treasures you must use a large searchcoil . . . the largest available for your detector. Twelve-inch diameter searchcoils are designed to be especially effective in cache hunting situations.

No matter which advertisements you may have read or which stories you have heard, the laws of physics are eternal. The larger searchcoil you use, the larger and deeper penetrating will be the electromagnetic field that your detector will generate. The three-foot electromagnetic field generated by your smaller searchcoil will not reach that four-foot cache, but the five-foot field of your 12-inch coil will.

As you may already know from your coin hunting experiences, the longer a coin has been buried, the easier it is to detect. Depending upon soil characteristics and other factors, freshly buried metallic objects can be detected to about one-half the depth of the same objects when buried for a longer time. This same phenomenon holds true in the detection of buried caches.

Caches come in all sizes, but most are small. The treasure you seek may be found in a small tobacco tin, but it may be contained in a big steamer trunk! Regardless, you must not take a chance; so, use a large searchcoil. There is no doubt that even the best treasure hunters have left deep caches that were beyond the range of the finest detectors available in earlier years. These caches await the hunters with 21st century instruments capable of finding them.

Many failures can be attributed to the hobbyist who is thoroughly familiar with the techniques of coin hunting but is inexperienced in cache hunting. Because he has full confidence in his detector to locate deep coins, he may overestimate its power when he begins to cache hunt. He may believe that since the cache is large, he has all the power needed to locate it.

So, he envisions himself a cache hunter and conducts proper research to develop a good lead at, say, an old church or mission site. It was used as a hideout after a robbery, and loot was buried there which has never been recovered. A great deal of time is obviously required by this research, and reaching the site may call for considerably more time plus expense, including the purchase of additional recovery equipment.

Finally, after this expenditure of time and money, the would-be cache hunter is on-site, ready to scan with his discrimination-type, coin-hunting detector that will leave our self-designated cache hunter almost helpless. Perhaps he'll be able to locate some shallow relics or old coins.

Occasionally, you can see the above scenario portrayed in treasure magazines. The article is accompanied by a picture of the individual(s) on site. Look at the searchcoils on their detectors. If the searchcoils are small, this hunt probably produced a few old coins or relics but probably not a cache. Certainly, not one that was deeply buried in mineralized soil.

Detectors Make a Difference

The authors have not intended to degrade any detector or manufacturer. Some instruments produced solely for coin hunting are highly satisfactory, but we don't recommend them for cache hunting. And, some inexpensive instruments can be adapted with larger searchcoils, but they are generally so unstable and so poorly designed that they are virtually worthless in the field.

Dealers may sometimes be faulted because they do not explain that a particular model does not possess deepseeking capabilities or that it lacks the versatility required for cache hunting. The popularity of certain instruments can become so widespread in an area that limitations of these detectors are completely overlooked.

Knowledgeable hunters certainly listen to manufacturers' and dealers' claims, but they depend primarily upon field test data – usually their own.

Professionals who depend on treasure hunting for their livelihood demand the best quality instruments and usually own two or more detectors on which they can rely.

Time in Research

Most cache hunters spend a major portion of their time in research, seldom mentioning their occupation except to another professional. Since proper research may require extensive travel, their expenses to find a single cache can be considerable – even before they turn on their detector. Sometimes, they pay sizeable sums to obtain information. Often, they agree to share the cache on a percentage basis, a common practice for gaining permission to search on private property. Occasionally, a special detector must be purchased because of the nature of the ground where a cache is sought. Proper financing, as well as patience, is required.

Cache hunters seek real treasure – financial wealth – and a bundle of it! Of course, not all are successful every time. The beginner should realize this and not become discouraged. We advise working on several projects simultaneously. Since research can be expensive and time consuming, it is good to "double-up" on the uses you can make of it. Always remember that there are literally millions of dollars stashed in the ground waiting to be found. If you persist, sooner or later you will hit a cache. It may be only a few hundred dollars tucked in a tobacco can; then again, you may become wealthy from pursuing this interesting occupation.

Techniques necessary for successful cache hunting differ somewhat from those used in scanning for other types of treasure. In searching for coins, for example, you generally used the Discriminate mode of your detector with occasional ventures into All Metal. For cache hunting, we suggest the opposite. In fact, to find caches most effectively, we recommend that you use the All Metal mode almost exclusively with no discrimination of any kind.

Let's say you are searching for a covered iron pot that is filled with gold coins. In such a situation your detector will signal to you about the big pot, not all those coins inside. If you were scanning with discrimination, your detector might reject the iron pot, and you'd never dig it. What a disaster! Use your All Metal mode, but remember what we've said earlier and so many other times – if you're to be *really successful,* you must be prepared to dig lots of junk.

Unless you're searching for a cache in a building – where you know that it cannot possibly be *too far* away – always use the largest searchcoil possible. Remember that larger searchcoils can detect larger objects deeper. Money caches have been found at all depths (arm's length seems to be the norm), but you want to be prepared for extremes. In some areas, where washing has occurred and drainage patterns have redesigned the landscape, caches have been found more deeply covered than when they were originally buried. All the more reason to use the larger searchcoils – even the Depth Multiplier Bloodhound!

The Depth Multiplier

The Depth Multiplier, made famous by successes of professional cache hunters, is manufactured by Garrett and is affectionately called the Bloodhound. It is a two-box super-deepseeking searchcoil manufactured for use with Garrett's computerized Grand Master Hunter and its ADS detectors. The Depth Multiplier is designed to detect only large objects and offers the greatest depth possible with a metal detector.

An important feature of the Depth Multiplier is that it will not detect small objects. In an old farmyard, for example, you won't be bothered by small trash littering the soil. You will be able to dig only larger targets, approximately quart-size and larger.

The Depth Multiplier attachment is easy to use in the detector's All Metal mode. Do not use the Discriminate mode. No ground balancing is required. Always wear headphones and adjust the audio threshold for faint sound. Be sure you aren't carrying a large metal object such as a shovel or large knife, even though a few coins in your pocket may not matter. Hold the detector with your arm extended and walk at a normal walking speed across the area you wish to search.

Listen carefully for an increase in the audio level. When you hear the louder sound, stop and scratch a mark on the ground with your shoe. Continue walking without adjusting any of the detector's controls. When you have walked across the object you have just detected, the audio will return to its threshold level. Continue a few more feet before turning around and walking a return path. At the point where the audio increases as you are walking from this different direction, make another mark on the ground. Your target will lie at the centerpoint between your two marks on the ground.

Successful searching for caches requires considerable experience . . . and thinking. You must learn to put yourself right in the shoes of the person who hid the cache for which you are searching. You know that that person didn't just run out of his house or jump off his horse haphazardly and dig a hole to bury a can full of money. He behaved as you would if burying a cache; you'd select a secret place and a secret time to bury it . . . perhaps, at night during a thunderstorm. And, your "secret place" would be one that you could find in a hurry!

Hide Your Own Cache!

Practice this yourself. Put some money (or something similar) in a glass jar. Bury it near your house. Would you do it in broad daylight? Would you just walk out into the yard and start digging? Probably not, because you wouldn't want anyone to see what you were doing. So, choose the right time and the right place to bury your cache. After you've done this, you'll be able to ask yourself the questions that probably occurred to that person who hid the cache you are seeking. Can I find it quickly? Could it be discovered accidentally? Will it be safe? Many other questions will come into your mind as you recover your own cache and relocate it a time or two. This is good experience that will make you a better cache hunter.

Be especially careful not to hurry when searching for caches. Scan carefully and cover every square foot of ground. Do not be in a rush; to do so may cheat you out of a valuable find.

When searching for a cache behind or inside a wall of a house, some discrimination is acceptable to reject nails. Even when using it, however, we recommend that you turn discrimination controls to the lowest setting possible. In either the All Metal or Discriminate mode you'll have more than enough sensitivity to detect almost any size cache in all walls, despite their thickness or type of construction.

When your treasure map leads you to a stucco wall containing a wire mesh, here are some tips to help you detect through that mesh. With your detector at minimum discrimination, place its searchcoil against the wall. Carefully slide the searchcoil over the wall which will lessen interefference from the mesh. You may hear a jumbled mass of sound, but listen for significant changes that could indicate you have located your cache.

Some prefer to search walls that are built with a mesh by holding the searchcoil several inches or even a foot away. Getting the searchcoil this far away should take care of the jumbled sound yet still let you detect large masses of metal such as a cache.

Professional cache hunters always make allowances for the condition of the search area and the fact that their cache may be deeper or smaller than anticipated. They take every precaution they can – beginning, of course, with the use of a deepseeking detector with the largest coil possible. When you are using the All Metal mode of a detector, excessively magnetic mineralized rocks can cause a problem. This is especially true when they are located near your target or near where you expect to find it. Known as "hot rocks" to electronic prospectors, these little pests are geological freaks – rocks that somehow have gotten themselves in the "wrong" location. Because they are unlike the soil and rocks in which they are found, they upset the ground balance of your detector and give a false positive signal.

Electronic prospectors have learned to use the features of modern detectors to deal quickly with these problem rocks, and so can you. When you suspect a "hot rock," merely switch to your Discriminate mode (using minimum discrimination). Check the target again exactly where you got the positive signal. If the audio decreases slightly or stops completely, you've encountered only a mineralized rock. If you continue to get a positive response, you should investigate this target.

Trying to use the discrimination controls to identify a target further is generally useless in cache hunting. Remember that your detector is responding to the container in which the cache is contained, and it is usually made of some ferrous (iron) substance. In special situations when you are looking for coins or precious metal in a non-ferrous container such as a glass jar or leather bag, you can increase discrimination until you are convinced the target is not iron. This situation seldom occurs, however, and we recommend that you investigate all positive signals. In using discrimination, your primary purpose is to determine only whether your target is metal . . . not what type of metal.

Never pass a suspected treasure site because you have been told that it has been worked before. More treasures are missed than recovered. In researching the novel, *The Secret of John Murrell's Vault,* one of the authors returned to the spot where he had once found only a deep hole instead of the treasure he expected. On re-inspecting that hole, he reflected that maybe the REAL treasure had been buried even deeper, with only a sampling of items left in a container above to satisfy anyone who might accidentally stumble upon this site!

Consider the old parks where coins continue to be found . . . and, not all new coins, either! These parks never seem to be completely hunted out. Now, consider the rugged, highly mineralized terrain where most caches are found and the eternal question of how deeply they were actually buried. These caches are far harder to find than coins. Remember, also, that anyone who searched a site in past years probably did so with a detector whose capabilities are far exceeded by your modern instrument.

Recovery Tools

We recommend a long steel probe that you can use to save time where soil conditions permit. If you believe that your detector's response indicates a target large and deep enough to conform to the cache for which you are searching, you can probe the spot before digging. Length of your probe will determine how deep you can search. Experienced operators recommend one at least 40 inches long. They have learned to probe carefully to determine just what kind of target they have discovered. Of course, you have a good idea of what you are looking for, and that helps.

You'll know easily if your probe hits a glass jar, or if it hits a junk piece of metal your probe can easily penetrate. If you find a tin can, the probe may penetrate it to let you know if something is inside. Depth at which the object is found can give you some idea when it was buried. Many cache hunters who use probes become so proficient with them that they can feel a newspaper when the probe passes through it. The real old-timers even claim to be able to *read* the newspaper with their probe!

We recommend that you build your own special probe that has a steel ball bearing welded near the point of the shaft. A one-half-inch ball bearing welded into a three-eighth-inch steel rod permits your probe to move up and down easily with no restrictions and lets you determine more easily just what you have found. The oversized bearing actually "digs" or creates the hole, letting the thinner rod slide in without friction.

A Low Profile

Most cache hunters try to avoid calling attention to themselves. One way to do this is by carrying detectors and all other equipment into the field inside a backpack. You then appear just as any other hiker. A large backpack will usually accommodate a 12-inch searchcoil and a Depth Multiplier attachment as well as small shovels, your detector's housing and the small tools necessary for an average recovery. There are numerous reasons for not calling attention to yourself or your search for caches. First of all, you don't need the attention of curiosity-seekers. Plus, if word ever gets out about your recovery of a cache, you'll be amazed at the number of people who claim rights to all or a part of it!

Never put your trust in a verbal agreement with a landowner and never leave an open hole after you have recovered something. The first admonition is self-evident. As the wise man once said, "A verbal agreement isn't worth the paper it's written on!" Concerning the second, let the other author relate *his* experience with an empty hole:

After executing a written agreement with a landowner, a cache hunter found and removed the treasure he was seeking. Unfortunately, he made the mistake of leaving a deep hole which was discovered by the landowner. Thoughts of that deep hole filled with gold probably flooded the landowner's mind, and he contacted his lawyer immediately, neglecting to mention the written agreement. The lawyer advised a quick civil suit for recovery and damages, and the cache hunter faced time in court and legal expenses even though he believed himself protected with the written agreement. To avoid trouble, he transferred ownership of his real property to a relative and let the lawyer fume since there was nothing on which he could file an attachment in court. The landowner gave up fighting a lost cause, and the issue was forgotten – by all except us cache hunters who learned a valuable lesson we won't forget!

When you are working with partners, make certain that all arrangements are made in writing before a cache is discovered. Many of us have had unpleasant experiences, particularly in working with non-professional treasure hunters, such as landowners. Generally, you can trust a cache hunter who makes his living in the business. He cannot afford to have his reputation clouded by a squabble over property rights. Plus, he has handled "found" money before and does not tend to get as excited about it as a non-professional. More on this subject can be found in Chapter 19, which includes a section on laws and a model Search Agreement for your consideration.

Taxes are also a subject covered in more detail in Chapter 19. The Federal Government demands its percentage of any income you derive from treasure just like that from an investment or salary. Similarly, states and municipalities that tax income demand their share. In summary, however, our advice is to pay what is due when you owe it. If you can prove you're in the treasure hunting business, proper expenses can be deducted. Tax requirements differ from state to state, but most of them generally seem to be trying to get more of the recoveries for which treasure hunters have risked time and money – and, sometimes, their lives.

So, our advice to you as a cache hunter is to keep a low profile, don't call attention to yourself, pay your legitimate taxes and insist on all your rights.

Because the new computerized detectors will search deeper and with more precise ground balance, we urge that you give them a chance to help you find the *big money* that's waiting for cache hunters.

Chapter 6

Searching Buildings and Cabins

The title of this chapter could well have been "Ghost Towning." Perhaps it should have been because this phase of treasure hunting is becoming more and more popular among metal detector hobbyists everywhere. The authors chose, however, not to try to lump together in a single chapter all the potential treasure hunting activities available in ghost towns. Truly, these locations are veritable wonderlands for a metal detector hobbyist. They encompass all phases of treasure hunting. Searching for caches, seeking coins, hunting for relics . . . all can be pursued profitably in a ghost town as well as the search of buildings, cabins and other structures that were once occupied.

What about searching houses where people still live? Provided they are old enough and have enough "history," occupied homes can present many targets. You can look for caches hidden and forgotten by previous dwellers. You can seek jewelry or silverware that was hidden for safekeeping and never recovered for some reason. You can find coins and other items of value that might have fallen through cracks to rest under floorboards or between walls.

Treasure waits to be found wherever men and women have been. Humans misplace, lose and hide items of value in locations where they can only be discovered by a modern metal detector. Of course, you should never fail to use your eyesight in searching any location. What can be discovered simply lying on the ground or on the floor of a structure is amazing. When you are searching an old cabin or house or hunting anywhere else in a ghost town, you should always remember that surface items were probably picked up by relic and antique hunters long, long ago. You will need a quality metal detector because most of the objects you seek are lying beneath the ground or are concealed from sight in some other way.

But, consider this true story. A treasure hunter was searching a Colorado ghost town. In one building he raised a window and there – just lying on the sill – was a $50 gold piece worth several thousand dollars.

Where to Search

Any place where people once lived or conducted business will produce treasure that can be located by a metal detector. Thousands of abandoned homesteads, stores and commercial establishments, schools and churches as well as townsites, forts and military installations await you. The list of places where people "used-to-be-but-no-longer-are" is truly endless.

And, many of these locations have never been searched! We have discussed more than once the importance of not being intimidated by the fact that a specific location has been searched before. Remember that metal detector capabilities have improved dramatically in just the past few years and that the proficiency of individuals can vary widely . . . even with the best and most modern instruments.

When someone tries to dissuade you from hunting anywhere by telling you that an area has already been searched, we suggest that you answer, "I've never searched it myself with *this* metal detector."

Perhaps you will find something that was overlooked previously because either the detector or its operator did not have the capabilities of you and your instrument. Perhaps it will be something very valuable!

Since research is first on the "to-do" list of any competent treasure hunter, you will want to find out specifically where to find ghost towns or buildings and cabins that can be searched. There are many books on this subject that can be studied at your local library or at the shop of your metal detector dealer. Historical societies and tourist bureaus are excellent sources for this information. Ghost town maps are sometimes available.

A Family Pastime

Searching old buildings and cabins in a ghost town is a pastime that you can combine with a family vacation. Perhaps your travels take you to unusual and interesting places . . . old abandoned towns and mining areas. Can you believe that thousands of tourists travel through such areas without stopping? Yes, even metal detector hobbyists who don't realize all that they are missing! Relics, old coins and other valuable items are just waiting to be found. In many of the areas that especially cater to tourists, you will find libraries and visitor facilities especially equipped to suggest places for you to search with a metal detector. These free sources of information are generally more accurate than hand-drawn maps and guides you receive from individuals.

Don't forget all the many things that you have learned about research, however! Use these techniques to the fullest. Remember how helpful old-timers can be and what you can learn from long-time residents. In many of the tourist areas, unfortunately, you will discover that local citizens are not particularly interested in strangers, especially if the strangers are not important to their livelihood. Some hobbyists become discouraged when they receive little or no information from local citizens or are treated almost rudely. Don't you give up! Good people exist, and there is always the library. Keep looking for "that spot" where you can find treasure with your metal detector. Persevere, and you will be successful.

Discrimination

Modern metal detectors are ideally suited for searching cabins and buildings, even those with thousands upon thousands of small nails. These small, iron objects made searching houses oh-so-difficult for many of the older detectors. You'll still encounter some problems, but the modern detector with discrimination permits you to search buildings quite easily. And, you can be certain that you've searched thoroughly!

When you're scanning inside a wooden structure with literally countless nails, you can expect your metal detector to respond with multliple target signals. Of course, you don't want to react to every signal—even the strong ones. You don't want to tear out a wall just to locate a large nail. We recommend, therefore, that you operate your detector in its Discriminate mode, using only enough discrimination to reject troublesome small nails. You'll not be likely to overlook a large money cache!

If your main goal is to detect small, coin-sized objects, your regular searchcoil approximately eight inches in diameter is a good choice. The smaller Super Sniper coil may be an even better choice. Because it is smaller, you can get it into tighter places. Too, with its narrower diameter the Super Sniper or smaller coil won't be reporting about as many nails in its electromagnetic field.

When scanning around window and door frames, be alert for signals you receive from the iron sashes that balance suspension of the window frames. Don't tear into a wall looking for treasure until you have exhausted all techniques for peering into that wall by other means. Most wall areas on older houses can be visually inspected simply by pulling slightly back on a single board and shining a flashlight into the cavity.

A cache that is still intact in an abandoned building will generally be some type of ferrous or metal container. We used to recommend that no discrimination be used in searching for such a prize, but the new detectors with precise controls have outmoded this advice. As indicated above, use just enough discrimination to get you past the nails. Even the cache in a small tobacco tin should generate a signal that you can hear.

Some treasure hunters use a little more discrimination, especially when searching around the large nails present in joists and rafters. Use one of these large nails for bench testing and increase the discrimination control of your detector until the nail is ignored. At this level of discrimination, set your audio threshold just above the silent level with only a faint sound and search with the searchcoil approximately four inches or so away from the walls and ceilings. Your audio response should remain fairly even and enable you to locate all ferrous (and non-ferrous) objects larger than the nail for which you have set your level of discrimination. If the sound is still too erratic or jumpy, decrease the sensitivity (depth) control to permit easier operation.

Too much discrimination could cause the detector to reject a tobacco tin, small can or iron box. Too little discrimination permits the detector to respond so loudly to small nails that you might miss larger valuable objects. Regardless of your discrimination level, a modern detector will respond to non-ferrous targets as small as a coin. Practice in your own home with various levels of discrimination.

Non-Ferrous Targets

When you're searching for such non-ferrous targets as a coin cache in a glass jar or earthen pot, different procedures are needed. You will want to use the All Metal and perhaps even the Discriminate modes of your detector. Using this dual mode method, you will search in the All Metal mode and switch to Discriminate to check out targets you encounter.

After setting your audio threshold, adjust your discrimination controls to reject large ferrous targets such as iron pipes, tin cans and large nails. Begin searching in the All Metal mode. When you have located a target, pinpoint it and move your searchcoil away slightly. Then, switch to the Discriminate mode and press the searchcoil against the wall directly over your suspected target.

When the target consists of coins, brass or some non-ferrous object, the sound will remain positive. Practice by locating a known iron target such as a window sash inside a wall. Place a non-ferrous target such as a piece of brass, aluminum or bag of coins nearby. Pinpoint both targets in your All Metal mode. Switch to Discriminate and practice until you can "reject" the iron but not the treasure.

Always make certain to let the searchcoil remain in direct contact with the wall surface when attempting to identify all targets.

When you slide the searchcoil over a wall to pinpoint targets, the searchcoil will sometimes come very close to or in direct contact with a nail. You will hear a small, sharp response, which you soon should be able to identify. Either ignore this response or pull your searchcoil back several inches and double check.

Sensitivity

Many of the early detector models, even with discrimination, could not be operated effectively among metallic targets such as these small building nails. Retuning was a continual problem. Modern instruments have eliminated this problem, even while increasing sensitivity. In the old days, we appreciated the sensitivity of these early VLF detectors, but we had to grit our teeth while getting the metallic responses. Sometimes, we'd say, "Well, you can turn a highly sensitive detector down, but you can't turn a low sensitivity detector up!"

Now, we can have our cake and eat it too; that is, have ultra sensitivity, along with discrimination that lets us detect most effectively with it!

Brick chimneys are familiar occurrences in many abandoned buildings, and treasures have been found behind their loose bricks. Remember, however, that most bricks are made of highly mineralized clay with iron in the conductive soil that was baked. Is this an impossible problem? Not for a modern metal detector with circuitry that lets you ground balance it precisely! Simply ground balance your detector against a chimney just as you would against the ground. Most likely you'll find no chimney stones or bricks that you cannot properly ground balance. Be alert, of course, to nearby metal or pipes within the chimney. You cannot ground balance your detector if metal is present.

You'll be surprised at how many old buildings and cabins you

Facing
A pocket scanner is ideal for searching old buildings like those at this ghost town. Its penetrating power scans deep within floors, walls and ceilings.

Over
Searching around a chimney at an old home should present no problems when the detector is properly ground balanced against minerals present in the bricks.

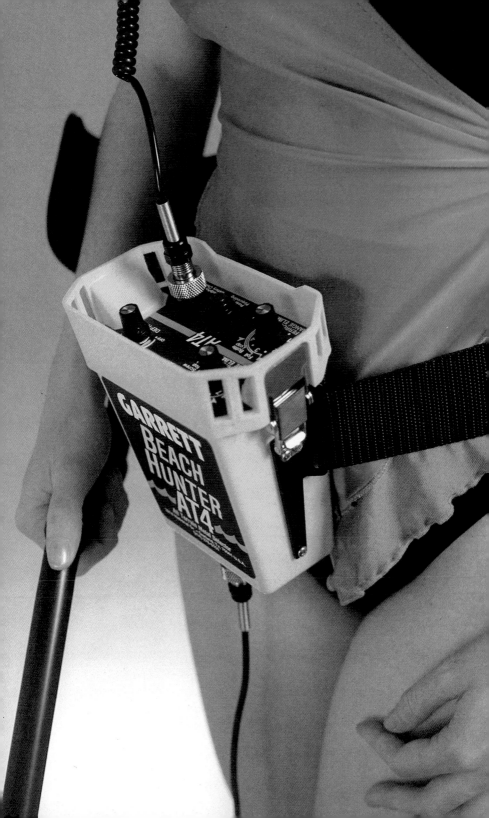

can find to search. Just a few points to remember: Never tear down or otherwise destroy old buildings. In fact, you should leave all structures in better condition than when you found them . . . without harm or defacement of any kind. Walk away from buildings and cabins that you have searched, leaving them in such condition that no one can really tell whether you found treasure there or not. Destroy nothing. Do not tear out any boards that you cannot replace easily. Use common courtesy at all times. Remember, you might want to return!

Facing Top
A variety of tools such as this long-handled scoop are used by water hunters for recovering prizes found in deep or murky waters of ocean surf and lakes.

Bottom
Some detectors such as this Garrett Sea Hunter are designed for searching under water. This instrument is guaranteed to operate to water depths of 200 feet.

Over
Garrett's AT4 Beach Hunter is designed to find teasure anywhere, but especially near the water where it can be worn on the hip out into the surf.

Chapter 7

Searching Beach and Surf

Welcome to the newest frontier of treasure hunting with a metal detector – the beaches and surfs of the world. Believe us when we tell you that there's wealth to be found near the water – and lots of it!

Yes, treasures are there right now, just waiting for you. And, new modern metal detectors make finding this seaside wealth as simple as locating new pennies lost in a park. At the water's edge, however, you'll be looking for expensive jewelry or relics from shipwrecks with today's detectors. These marvelous instruments are designed to overcome both the mineralization of beach sands (terrific in some areas) and the effect of wetted salt (always bad) in ocean water. There was no way – even just a few years ago – that the most modern detectors could have been used under such conditions. Yes, today's detectors have truly opened a new frontier. Come and explore it!

Over two thirds of the earth's total surface – nearly 200 million square miles – is water. Since the dawn of creation, man has lived at the water's edge or very near it. Commerce, transportation, recreation, exploration, warfare and the search for food have compelled men and women to return to water whenever they have strayed. And, you can be certain that whenever man made contact with water, he brought wealth with him . . . and, that he inevitably left some of it behind.

These possessions and other lost things of value can be found today in many forms. Perhaps it is a simple keepsake or talisman, a single coin or barter item, a trinket or crude jewel; then again, the wealth awaiting you on the beach and surf is sometimes a cache of doubloons hidden away by pirates or conquistadors centuries ago or the cargo of a ship that foundered in a storm. The contents of Davy Jones' locker, which includes all lakes and streams as well as the ocean, are beyond belief. We feel confident that if all this wealth could somehow to be evenly distributed over the earth, every man, woman and child living on it would be able to live comfortably for the rest of his or her life.

The world's oceans, lakes and streams, therefore, offer vast storehouses of lost wealth that await the treasure hunter equipped with a modern metal detector. Beaches at the entrance to Davy Jones' locker present the most accessible areas for hobbyists to begin searching. These vast sandy stretches are attractive to the metal detector hobbyist for still another reason. As rules and regulations for using detectors have grown in complexity, public beaches and the waters they touch become even more appealing as a recreational location, a site for pursuing this great pastime of treasure hunting.

From the very first time a hobbyist turns on a metal detector he or she is advised to seek treasure where treasure has been lost – in other words, where people are known to have lived and played. Thus, it has always been the continuing goal of hobbyists to search where targets are most plentiful . . . to seek treasure where it is hiding. You can believe us when tell you that it has been our experience over decades of treasure hunting that busy beaches generally will yield treasure more valuable and in greater quantities than just about any other site.

The 'Blanket' of Treasure

Think of all the coins, jewelry and other valuable objects that fall into the sand. While you scan a metal detector over the beach or ocean bottom, constantly keep *this* vision in mind: only a few feet beneath the sand's surface a veritable *"blanket"* of *treasure* awaits the treasure hunter. And, this blanket is continually being replenished! Unless someone quickly recovers it or it is found by a metal detector, every object that falls on the beach gradually works its way down to this blanket of wealth lying on bedrock or the most shallow clay strata.

Where people congregate, treasure can be found. There can be no disputing that statement; it's that simple. Try this test.

Visit any local park on a pleasant spring or summer day. Count the people and watch their activity. How many did you count? Chances are that you saw a few dozen. What were they doing? They were probably walking, picnicking or perhaps engaged in some sports activity.

Now, drive to a local swimming beach. Make the same observations. How many did you count and what were they doing? You probably counted the same few dozen, plus several hundred more who could lose valuable treasure. And, they too were walking, picnicking or engaged in some sports activity. But, their frolicking and horseplay in the surf or dunes seemed far more likely to dislodge jewelry and other treasures than the sedate activities of a park.

You can be sure that treasure will be lost at that beach every day. And, I don't mean "cheap" treasure. People consistently wear expensive jewelry while sunning or swimming. They either forget they have it on, or they don't understand how they could lose it. It can't happen to me, they must think. But, it will . . . and does!

Beach Treasures

Treasures awaiting your metal detector on the beach and in the surf include coins, rings, watches, necklaces, chains, bracelets and anklets, religious medallions and crucifixes, toys, knives, cigarette cases and lighters, sunshades, keys, relics, bottles, fishnet balls, ships' cargo and other items that will soon fill huge containers. And, for some lucky, persistent and talented hunters, their dream will come true. They will indeed find that chest of treasure hidden by a buccaneer or 17th century pirate who never returned to claim his cache.

It's hard to understand why people wear jewelry to the beach. Yet, they do, and they often forget . . . even about valuable heirlooms and diamond rings. But, whether sun bathers and swimmers care about losing their possessions or not, it's just the same for the beachcomber. All rings expand in the heat; everyone's fingers wrinkle and shrivel in the water and suntan oils merely hasten the inevitable losses. Beachgoers play ball, throw frisbees and engage in horseplay. These activities fling rings off of fingers and cause clasps on necklaces, bracelets and chains to break. Into the sand drop valuables where they quickly sink out of sight to be lost to all save the metal detector.

How many times have you watched coins, jewelry, keys and other beach "necessities" being placed oh-so-carefully on the edge of a towel or blanket? Then, in a hurry to escape a sudden storm or just through carelessness, the sunbather grabs and shakes the blanket. There go those "necessities" into the sand. Even though the valuables are sometimes immediately recovered, many are never found except by a metal detector.

Boys and girls play in the sand. Holes are dug, and sand is piled up and made into castles and other elaborate structures. In this process toys, coins, digging tools, jewelry, knives and other possessions are lost until the metal detector or keen observer discovers them.

Beachcombing treasure hunters have been blessed by modern detectors. They are ideal for the beach and surf. Today's automated detectors with discrimination are perfectly suited to search heavily mineralized sand and in salt water. These instruments ignore iron magnetite (black sand) and salt minerals, and they permit discrimination to be adjusted according to existing beach conditions. Automated models can be operated equally well from zero discrimination through pulltab rejection. Some instruments such as Garrett's AT4 Beach Hunter are designed to be taken right into the surf where a little splash or submersion won't bother them.

Submersible Detectors

Several detector companies build instruments designed to function under the water. While the AT4 (and Garrett's famed underwater Sea Hunter) are designed to be submersible, many other detector models are not. Make certain you are absolutely certain about the environmental protection guaranteed by your particular instrument. Any kind of water can ruin the circuitry of a detector, but salt water is particularly harmful. If your detector is not protected by its design characteristics, you must be careful at all times. Protect your instrument not only from submersion but against being splashed by a casual wave while you are searching or when you have laid it down on the beach.

Many searchcoils are submersible. All of those manufactured

by Garrett can be submerged to the connector; but, always check with the manufacturer if in doubt about any aspect of your instrument. Be careful of water that may enter the detector stem; not all units have a plug to prevent water from running into the control housing. To be safe, immediately after using a detector in water, drain the lower stem. If water is not drained, it may flood the instrument the first time its searchcoil is placed higher than the control box.

For greatest success on the beach always use headphones. Of course, most veteran treasure hunters use headphones no matter where they are searching. They are especially necessary on the beach where wind, surf and "people" noise will mask detector signals and cause many good targets to be missed. Any type headphone is better than none at all, but the best are those with ear cushions and adjustable volume controls. Coiled cords are recommended along with right-angle plugs.

Since large cushioned headphones can become hot and uncomfortable, smaller versions are available. Even though these lighter models with small ear coverings do not mask out as much noise interference, they can be used effectively.

How to Search

Essentially the same rules that apply to your metal detector for coin hunting will apply to beach hunting with the modern instruments. You need have no fear of mineralization in the sand or the effects of wetted salt in the water. Many experts will advise zero discrimination at all times, following their rule of "Dig all targets." We bend that rule a little sometimes and suggest a little discrimination, especially when the beach is especially littered with pulltabs and other trash – as so many of them are.

Of course, on your beach with lots of trash, the small Super Sniper searchcoil can be especially effective. Not only can it get into tight spots but it will never find as much trash because it just doesn't have as many targets in its smaller electromagnetic field.

Always pay close attention to what your detector is telling you – that's one of the reasons for headphones. This advice is especially pertinent on the beach when multiple signals are common.

Listen closely and develop the experience to know that your detector is signalling. It will always report what is under its searchcoil. Your detector will never lie, but you can misinterpret its signals.

When you are using the new, modern detectors for beach hunting you will soon find that there several major differences between coin hunting and searching the water's edge and surf:

- Where to search;
- What to look for and expect to find;
- How to recover your targets;
- Dealing with the water itself.

Where to Search

Dealing with *where* probably tends to separate the beginners from the experts more than any other factor. The new, modern detectors have served as great equalizers in ability. Still, knowing how to use the tides, the wind and the weather generally can help locate those areas where treasure is more likely to be found.

Of course, the dedicated treasure hunter always has an answer to the question of "Where?" That answer is "research." And, it is as true on the beach as anywhere else. Beyond that, experience must be the teacher. Inquiring and attentive hobbyists continually pick up ideas from other more veteran beachcombers, but the final decisions must be based on individual perceptions and intuition. Experience alone will educate the beach hunter about places that never produce and other places that are often rewarding. A knowledge of storm, wind and wave action will often rescue someone who is studying a new beach. Our books and those of other treasure hunters list numerous research sources where both general and specific leads can be found for searching beaches with a metal detector. As the hobbyist researches these various sources, techniques and abilities will improve.

Always begin locally; your home territory is the area you know best. Use every source of leads and information; seek out old timers; visit or write chambers of commerce and tourist bureaus. Don't forget to contact historical societies; leave no source untouched in your investigation of an area. To speed up work always be specific in screening information concerning swimming beaches, resorts and recreational areas. Don't overlook the favorite beaches of years gone by, either. Also, ghost towns are not limited to mountainous areas; they can be found on beaches as well. Treasures from the past are always found in and around them.

When checking newspapers, pay particular attention to accident reports that will usually give the location or at least the name of a particular beach. Review old newspapers; be especially alert for the Sunday weekend or recreation columns that proclaim the holiday joys of swimming and sunbathing at local beaches. Advertisments of beachwear occasionally offer clues to areas of activity.

Don't overlook old postcards; antique shops can be a good source. If there is a postcard collector in the area, pay him or her a visit. Old picture postcards can be reliable X-marks-the-spot waybills to treasure.

But, don't be content to work only local beaches. Broaden your scope; it may pay rewards. For example, if you live on the West Coast, make a study of the history of such busy shipping areas as San Francisco Bay and Puget Sound. Many ships have gone down there, losing valuable cargos of silver and gold, much of which has not been found. Violent storms often churn up ocean bottoms and cast sunken treasure on the beach. Other estuaries and harbors may not yield the precious metals of the Golden Gate, but historical study of any coastal area can often reveal locations for profitable metal detecting.

Never overlook the possibility of finding flotsam and jetsam washing ashore from offshore shipwrecks. Regardless of the age of a wreck, some cargo – especially gold, silver, copper and bronze objects – will probably remain in fair to excellent condition for years, decades or even centuries. Gulf Coast and Caribbean shipwreck locations still yield silver and gold from the mines of Mexico and Peru. Gold and silver from California and other western states can be found along the Pacific coast. Often, rich cargo from shipwrecks is located.

When researching reports of shipwrecks, don't overlook Coast Guard and Life Saving Service records. Newspaper files and local and state histories are good sources of information.

Insurance companies and Lloyd's Register may provide precisely the data you need.

Assateague Island, off the coast of Maryland and Virginia, has proven to be the depository of much cargo from shipwrecks of yesteryear. Treasure hunters, scanning the beaches with their metal detectors, have found valuable coins and relics, some of which "marked" the location of larger treasures. Although much of the island is controlled by the National Seashore Service, portions are open to the public. Permission to search with your metal detector can sometimes be obtained on National Seashores; it never hurts to ask for permission.

Winds, Tides, Weather

Stay alert to current weather and beach conditions. You'll want to search at low tides – the lower the better. After storms come ashore, head for the beach. When oil spills deposit tar and oil on beaches, there's a good possibility bulldozer and other equipment used to remove it can get you much closer to treasure. Watch for beach development work. When pipelines are being laid and when seawalls, breakwaters and piers are being constructed, work these areas of excavation.

Wouldn't it be great if the ocean suddenly receded several feet, leaving your favorite hunting beach *high and dry?* You could walk right out and recover lost treasure so much more easily. Well, the ocean does recede slightly every day during low tide. About twice a day a full tide cycle occurs -- two high and two low tides. It's low tide that interests the treasure hunter because a drop of only a few inches in tide level can take the ocean several yards farther out, especially on gently sloping swimming beaches. This exposes more beach to be searched and also makes more shallow surf area available.

You can learn when maximum low tides occur by reading tide tables in newspapers or obtaining them from scuba shops or fishing tackle stores. Weathermen on radio and television in coastal cities often report times of high and low tides. On some days, especially after a new or full moon, tides will be lower than usual. Take advantage of these opportunities.

The successful beach or surf hunter begins working at least two hours before low tide and continues that long after the ocean begins to rise. That's four to six hours of improved hunting. Be alert to lowest or ebb tides when you can work beach areas not normally exposed. Timing search periods is important. A good strategy is to scan dry beaches during high tides and then follow the tide out, working a parallel path hugging the water's edge. Each return path is nearly parallel to the preceding one. When the length of the paths is too long, each path will veer outward as the water recedes. Wide searchcoil sweeps can offset these veering paths.

Listen regularly to weather reports and forecasts to learn of prevailing winds. Strong offshore (outgoing) winds will lower the water level and reduce size and force of breakers. Such offshore winds also spread out sand at the water's edge, reducing the amount that lies over the blanket of treasure. On the other hand, incoming wind and waves tend to pile sand up, causing it to increase in depth. Pay attention to winds and tides, especially during storms.

Weather is a contributing factor to tide levels, which can be dramatically altered by storms and high winds. A big blow moving in from sea may raise normal tides by several feet. When this occurs, wave action can become so violent that it is impossible and dangerous to hunt – even far up on the beach. But, the stage is set, and you should hit the beach when calm returns.

Conversely, an outgoing storm can cause lower tides and a compression of wave heights. These conditions and the changes they cause is a continuing process that controls sand deposits on the beach and in shallow water.

Storms often transfer treasure from deep water vaults to shallower locations. Plan a beach search immediately following a squall. If you are hardy enough, try working during the storm itself. It may be revealing. As the rain and surf hit the beach, gullies may form before your eyes, giving you an immediate path to the *blanket of treasure* lying at bedrock level. Always remember that such storms or other extremes in weather, wind and tides can make unproductive beaches suddenly become productive.

Sand Formations

Another reason for working beaches immediately after a

storm is that the beach continually reshapes and protects itself. Sands shift normally to straighten the beachfront and present the least possible shoreline to the sea's continuous onslaught. During storms, beach levels decrease as sand washes out to form underwater bars which blunt the destructive force of oncoming waves. Following the storm, waves return this sand to the beach.

To understand how articles continually move around in the shallow ocean, consider the action of waves upon sand. At the water's edge, particles of sand form the sand bank When a wave comes in, sudden immersion in water causes the grains of sand to "lighten" and become more or less suspended in the water. Such constant churning keeps particles afloat until the next wave comes in. The floating particles are then carried some distance by the force of the water.

In the same manner, coins, jewelry, sea shells and debris are continually relocated, generally in the direction of prevailing wind and waves. As they move, waves and wind carry material until a spot is reached where force of the water lessens. Heavy objects fall out and become concentrated in *nature's traps*. So, whenever you find a concentration of sea shells, gravel, flotsam, driftwood and other debris, work these areas with your metal detector.

Similarly, look for tidal pools and long, water-filled depressions on the beach. Any areas holding water should be investigated since these low spots put you closer to the blanket of treasure. As the tide recedes, watch for streams draining back into the ocean. These will locate low areas where you can get your searchcoil closer to the treasure.

As experience accumulates, you will discover "mislocated" treasure in areas away from people. How did this happen? Perhaps this is where people used to congregate; it might once have been a swimming beach. Then, for some reason, the old beach was abandoned along with its treasure. Another reason is natural erosion that redeposits objects. Even though such action is seldom permanent, always keep in mind the forces that cause it to happen – and, watch for them in action. These forces do not occur accidentally, and they can create treasure vaults for you to find and unload.

Pay attention the next time you get close to turbulent surf. When a wave breaks near the beach, notice that water appears brown because of suspended sand. Crashing waves transport this sand onto the beach. If the waves are breaking perpendicularly – at a ninety-degree angle to the beach front – most of this sand is washed right back out to sea by the receding water.

Waves rarely break perpendicularly; rather, they bring sand in at an angle that sets up a current. This angle of transport washes sand to either side of its origination point. Some of the displaced sand remains on the beach and some is washed out to a new location. The result of this action is sand movement in the general direction of the waves.

Understanding this phenomenon is important because the same "ocean transport system" via storms and high waves causes a redistribution of treasure from the point where it was lost to its present location awaiting your metal detector. The ability of water to move heavier-than-sand material depends on its speed. Large waves and fast-moving currents can carry sand, coins and jewelry along a continuous path. When wave action slows down, movement subsides. When wave action picks up, movement resumes. Growing shores are "nourished" by material that has been eroded from a nearby stretch of beach. Heavy treasure takes the path of least resistance, moving along the lowest points of cuts and other eroded areas. As coins and jewelry are swept into new beach areas, they become fill along with the new sand. Being heavier, they gradually sink to lower levels and become covered. When a beach or shore area has become fully "nourished," the buildup essentially stops, leaving the treasure buried and awaiting the signal of a metal detector.

Since shorelines and beaches are continually being reshaped, you must be observant. One key to success is establishing permanent tide and sand markers. Such markers can be a piling or structure readily visible at any time. Ideally, your water marker will be somewhat submerged during both high and low tides. Checking this marker lets you measure water depth at all times to learn if the water is rising or falling.

Your marker in the sand is important because it is a gauge of sand height. Chances of finding treasure increase as more of this marker is exposed, indicating low sand levels. There are high and low sand formations. At low levels you may find treasure as it becomes uncovered by the action of wind and waves. High formations do you no good except to serve as gauges when storms erode cliffs. Imaginative beachcombers expect such erosion to reveal accumulations of debris and treasure that have been buried for decades. As noted earlier, stay alert for references to old settlements or ghost towns. What has been covered for generations may be uncovered before your eyes.

Grid Searching

When searching a large area of beach or surf, you should clearly define your area of search and systematically scan every square foot. There are many grid methods to use, some simple, some elaborate. The simplest, perhaps, is to guide on your previous tracks as you double back and forth. Using a stick or other object you can draw squares in the sand. Work the first square completely and then draw an adjoining square and work it. These methods work if others don't destroy your tracks and lines as fast as you make them.

You can drive stakes into the ground or just guide yourself on piers, trash containers, trees and other permanent objects. Setting up and working grids is somewhat more difficult in the surf, but you can accomplish it if you try!

When not following the tide out, some hobbyists prefer to walk a path parallel to the water's edge – either on the beach or in the surf. They then turn around, move about two feet and walk a return path. Others prefer to start at the high water mark and scan down to the water. They then turn around and walk a return path about two feet to the side of their first path. This second method has more merit since it permits treasure troughs to be spotted more quickly.

Treasure Troughs

These troughs, or "cut" areas that bring you closer to bedrock and the treasure blanket lying there, sometimes form parallel to the water line. When the tide goes out, the troughs fill with sand; still, they can sometimes be found. While scanning a path between high tide and the water's edge, each time a find is made, either mark the location or remember it. After you have scanned some distance down the beach and made several finds, look back and study where you have worked. Observe the location of your finds to see if a pattern is developing. Most may have occurred in a narrow belt parallel to the waterline. If so, you've probably discovered the location of a buried trough where a storm or other wind and wave action have created a treasure vault. *Empty it!*

When selecting a beach on which to walk your grid pattern, seek one where you earlier observed a cut forming perpendicular to the waterline. High tides or waves pouring back into the ocean form these cuts, usually at low spots that have resulted from previous storms. Remember, cuts are important to you because they bring you closer to treasure; also, coins and jewelry washing off a beach are pushed into these cuts by streams of draining water.

Now that you will be searching in patterns rather than randomly, you will soon see the value of keeping precise logs of your treasure finds. Even with others working the same beach, it is likely that valuable patterns will emerge on the pages of your notebook that will lead you directly to "hot spots." You may think that with others and yourself steadily working a particular beach, all its treasure would soon be recovered. Why keep track, you may ask. You'll learn, however, that active beaches are continually *replenished* by "new" lost treasure and that all beaches add "old" treasures that tempests have withdrawn from storerooms hidden in deep water.

Scanning Tips

Do not race across the sand or splash around in the surf with your searchcoil waving in front of you. *Slow down!* Work methodically in a pre-planned pattern. Unless you seek only shallow, recently lost treasure, reduce scan speed to about one foot per second. Let the searchcoil just skim the sands and keep it level throughout the length of a sweep. Overlap each sweep by advancing your searchcoil about one-half its diameter.

Always scan in a straight line. This improves your ability to maintain correct and uniform searchcoil height, helps eliminate the "upswing" at the end of each sweep and improves your ability to overlap in a uniform manner, thus minimizing skips. Practice this method; you'll soon come to love it – and, especially its results.

We've touched on them elsewhere in this book, but it would be well to mention "hot rocks" here as well. Gravel on the beach may sometimes include pieces with enough mineral content to be classified as a detectable hot rock. Your detector will occasionally get a good reading on one of these rocks and sound off with a "metal" signal. When this happens, set your discrimination to zero rejection and scan back over the gravel. If it is a hot rock, your detector will ignore it, or the sound level will decrease slightly. The subject of hot rocks is covered more fully in Chapter 10, Electronic Prospecting.

Don't ignore either very loud or very faint detector signals. Always try to determine the source. If a loud signal seems to come from a can or other large object, remove it and scan the spot again. When you hear a very faint signal, scoop out some sand to get your searchcoil closer to the target and scan again. If the signal has disappeared, scan the sand you scooped out –

Facing
The beauty of a lonely beach is but one of the joys that await a treasure hunter who seeks the prizes that can be found here beneath the sand and in the water.

Over
The **blanket of treasure** awaiting hobbyists under beach sands and in the surf at the doorway to Davy Jones' locker contains wealth of all kinds.

you may have detected a very small target. It might be only a BB, but at least you'll know what caused the signal.

Remember. Your metal detector will *never lie to you*. When it gives a signal, something is there.

During your search near the water or in shallow surf, when you begin detecting trash (pulltabs?) in a line parallel to the waterline, search for a nearby parallel trough. Remember that more than one trough may have been created, and that those farther out can contain heavier treasure items. Walk out from your "trash trough" and seek out one that produces keepers.

When pinpointing, always try to be precise. Good pinpointing saves time and lessens the possibility of damaging your finds when you dig.

Various pinpointing and retrieving ideas and methods have been reviewed. Here are a couple of final suggestions that could increase your take. As part of your beach gear, consider adding a garden rake. When you encounter debris, seaweed and other materials spread over an area you want to scan, use the rake to remove such material. Try to place raked materials where they can be picked up by beach cleaners and not washed back out to sea. Removal of any overburden will let you scan your searchcoil closer to the ground, moving you down just a little closer to those deep treasures.

Facing
This prize is a 16th century Spanish religious icon found by Charles Garrett with a metal detector in the surf of Massacre Beach on the island of Guadeloupe.

Over
A metal detector especially designed for beach hunting offers healthy outdoor exercise as well as the thrill of discovery and financial rewards.

Try exploratory trenching to clean out deep troughs and locate glory holes. Choose a spot where you have found a concentration of good objects (not items flung from a blanket) and dig (as long as it's legal) a trench about a foot deep. This trench should be wide enough for you to insert your searchcoil in its normal scanning position. Be sure to scan the sand you dig out. The trench length can be as long as you like. If you are digging in a spot where you have found several items close together, determine if you are in a natural drainage pattern. If so, dig toward the low side (the direction water flows) because that is the direction that coins and jewelry will have been washed. If your trench is being dug to locate a trough, dig in a perpendicular direction to the "line" along which you were locating targets. You may have to dig several parallel trenches to locate the trough.

No matter, where or how much you dig . . . always fill your trenches and remove trash that you uncover.

Recovery Equipment

You'll discover a big difference in treasure hunting on the beach and, especially, in the surf when it comes time to recover the target you have located. Because almost any type of digger can be used in loose sands, beginning beach hunters often lose sight of the importance of tools. Why, some even use their hands. We strongly counsel against this for several reasons, the first and foremost of these being the abundance of broken glass. In fact, we always recommend gloves, at least for the hand that does any digging.

Another reason for not depending upon hands as a digging tool is that the beach hunter cannot always expect to find targets in soft beach sand. We suggest two types of diggers: a heavy-duty garden trowel and a light-weight pick with a flat blade on one end. Just a quick whack with the pick, and you'll usually have your treasure. Of course, pinpointing is essential before digging is attempted. The hobbyist should begin with a trowel or small shovel and graduate to a pick-type digger with a long handle when pinpointing improves. The long handle permits uncovering targets without having to kneel on the ground.

Scoops are reasonably good in dry, loose sand. A quick scoop, a few shakes and there's the find. In wet sand, however, scoops are just a waste of time. It takes too long to work damp sand out of a scoop, except in the water where onrushing surf can help clean wet sand from the scoop.

Occasionally, a strong, thin digger-like screwdriver is needed since your finds may sometimes be buried in roots beneath trees and tree stumps. Digging can become difficult within a complicated root structure, and a strong, thin rod is needed to loosen the soil and make a hole from which the find can be recovered.

Storing Your Finds

Other gear needed for beach hunting includes an assortment of pouches, a secure pocket for storing especially good finds and a place for personal items. If you've hunted for treasure at all, you probably already have some ideas about recovery pouches. Let me offer just a couple of suggestions for the beach:

– Place *all* detected items in a pouch; carefully inspect your finds occasionally and discard trash properly.

– Use care in handling rings with stones. Often, mountings corrode during exposure. Examine jewelry with your pocket magnifier; when a mounting shows corrosion, handle that ring with extra caution.

– A fastener on a pouch is not a necessity on the beach unless you lay your pouch down carelessly or let it bounce around in your car. In the surf your pouch must be fastened, especially if you let it become submerged.

– Pouches for beach use should be waterproof to prevent soiling your clothes and sturdy enough to hold plenty of weight.

Concerning clothing, the best advice is to dress comfortably. But, protect yourself against the elements you're sure to encounter on the beach. Obviously, you'll want to keep warm in the winter and cool in the summer, but I caution you to shade exposed skin areas to protect against sun and wind burn. In warm weather shorts or lightweight trousers are recommended, along with a light (but long-sleeved) shirt, socks and comfortable shoes or sneakers. Don't forget about a hat. It should be wide-brimmed, preferably with some sort of neck shield. Even when you plan to hunt only on the beach, always be prepared to get wet. Sometimes an attractive low place in the sand will be yielding recoveries, and you'll want to follow it right into the water.

For the Surf

Equipment that you will need in the surf will, of course, depend upon whether you work in shallow or deep water. We can define shallow surf as water depth that permits you to dig with your hand tool, a scoop or some sort of digger – in other words, at arm's length. Deep water surfing is hunting in depths of about five feet, or the maximum depth you can safely wade in the water without swimming or floating.

Your choice of retrieving tools will depend upon soil conditions and personal preferences. In sandy areas, a scoop is fast. If the soil is muddy or made of hardened clay, you will need some kind of digger. In deep water, a long-handled scoop is required to retrieve your finds.

When the water grows colder, hip or chest-high waders and suitable underclothing will keep you warm and dry. When wearing waders, be alert or else you may bend too far. Suddenly, you'll find yourself wearing "convertible" gear: Your waders have been converted into a wet suit! Wearing a waist or chest belt over waders can reduce the amount of water that comes in. No matter what type of treasure pouch or pouches you use, they must close tightly. Water-hunting treasure pouches must have a secure flap covering. Some surfers use a sturdy open-weave bag or pouch with zipper or drawstring. Whatever equipment your ingenuity comes up with, keep it in good shape. Don't lose valuables through holes! It's a good idea to have several pockets that let you separate treasure and trash. But, whatever you do, never discard trash without carefully examining every piece. You may have inadvertently placed a good find in the trash pocket. Also, that item that looked so corroded and unrecognizable may turn out to be a valuable object. When in doubt about any find, take it home for closer examination, even an electrolytic bath for cleaning.

Some surfers use a flotation screen; others do not. If you use a converted land metal detector, you'll need a flotation device unless you mount the detector control housing on your body, or on the end of a very long searchcoil stem. The flotation device, of which there are several designs, is constructed with a one-half inch sturdy chicken wire screen. The screen opening should not allow a U.S. dime to pass through diagonally. If you are searching for smaller objects, the screen opening should be smaller. If your flotation device is large or contains your detector's control housing, the screen portion should hinge to permit rapid dumping of accumulated trash. The float can have recesses for the detector, a water bottle, your lunch and, perhaps, an extra tool or other necessity. Select a tube from the various automobile, motorcycle or bicycle sizes available. Since you are really not supporting much weight, the tube does not have to be highway bus or truck size. You should position the screen so that its bottom surface is about one or two inches below the water line. This facilitates quick washing of debris, mud and sand. In fact, when you dump several scoops into a well-designed flotation device, surf action should quickly clean the material.

Some hunters have said they prefer to place several scoops of dug material into their screen before they examine its contents and retrieve their finds. This can be efficient if you use a zippered bag which takes extra time to open and close each time you store a find. If the bottom is heavy silt and mud, it also might be quicker to dissolve and inspect a large of amount of soil rather than stir through your smaller retrieving scoop each time you make a dig.

An innertube can be punctured accidentally. This is a problem only if your detector housing is mounted on the float. Even then, I think you would hear the air escaping and you could rescue the detector before it got a bath. Placing your detector on a float is not recommended, however, for several other reasons. It has no protection from rain, the occasional wave or water splashed by swimmers.

It's Different!

In closing, we emphasize again to you that treasure hunting on the beach or in the surf is no more difficult than searching anywhere else. But, there are important differences! For example, we recommend that you schedule your beachcombing expeditions according to current (hourly) weather reports. Stay alert to weather forecasts and go prepared to withstand the worst.

Plan your treasure hunting expedition. Make a list of all you will need the day *before* you make the trip and check all gear carefully before you leave.

Always put batteries at the head of your list (see above). And, always check your batteries first if your detector should stop working. Some hobbyists take these longer life batteries for granted and expect them to last forever. Believe me, they won't. You'd be amazed at how many broken detectors can be "repaired" with new batteries.

Take along a friend, if possible. If you go alone, leave word where you'll be. Always carry identification that includes one or more telephone numbers or persons to call (with a quarter for the pay phone taped to the list). Your personal doctor's name should be on this list.

If there are no regulations to the contrary, you may want to search among crowds. But, don't annoy anyone. Angering the wrong person can result in immediate trouble, or you may find a complaint filed against you personally and the metal detector fraternity in general. You certainly wouldn't want to cause a beach to be put off limits for metal detecting.

Whenever possible, return any find to its owner. Try to oblige when someone asks your help in recovering a lost article. It might be feasible for you to loan them your detector and teach them how to use it. Who knows? You might add a new member to our brotherhood. When helping look for a lost article, it's a good idea to keep its owner close by throughout the search so that they will know whether you succeed or not. If you can't find the article, get their name or address; you might find it another day.

Do not enter posted or "No trespassing" beaches without obtaining permission. Even in states where you are certain that all beaches are open to the public, do not search fenced or posted areas without permission. Never argue with a "loaded shotgun;" leave such property owners to themselves.

Finally, remember that a modern metal detector is a wonderful scientific instrument. It searches beneath the sand, where you cannot see. It is always vigilant about the presence of metal. But, no detector can "do it all." You must develop powers of observation that keep you attentive to what a detector cannot see. Watch for the unusual! Sometimes you'll visually locate money, marketable sea shells or other valuables. The real benefit of developing keen powers of observation, however, is to enable you to enjoy the glories of the beach to their fullest and never to overlook the signposts pointing to detectable treasure.

As you scan along the waterline and observe the sands under the water, you may eye a coin shining in the water. Check the spot with your detector. Perhaps you found only a freshly dropped coin, or it could be the top layer of much greater treasure. And, how about that rock outcropping, the gravel or shells peeking through the sand, that accumulation of debris . . . any of these might mark the location of a glory hole. Remain alert and be rewarded!

There's treasure to be found near the water! And, vast amounts are waiting . . . enough for all. We sincerely hope that you'll join the rest of us beachcombers in searching for this lost and hidden wealth.

Chapter 8

Relics and Battlefields

Searching battlefields and other areas for lost relics is considerably different from cache hunting. Perhaps one difference is that the targets sought by relic hunters are usually quite smaller, such as a single button or a bullet. The techniques of scanning and locating, however, remain quite similar.

In the literal sense of the word *battlefield relics* are scarce indeed in the United States. Since the Civil War of 1861-65, our great nation has been blessedly free of wars fought on its soil. Even before then, the only actual war "battles" fought in what is now the United States were those in the East Coast states during the American Revolution and the War of 1812 and a skirmish or two in Texas during the Mexican War.

But, "war battlefields" notwithstanding, there are relics aplenty to be found. The richest trove of all, of course, is located in the Southern states where so many actions large and small were fought between Union and Confederate units. Fights with Indians left vast quantities of relics to be found throughout what was once frontier country. In addition to these obvious "battles" were many other skirmishes and actions of arms that resulted in relics to be found by today's metal detectors.

Incidentally, when you're searching "Out West," don't neglect antique barbed wire that is found by your detector. Just a single strand of old fencing can give you a real history lesson when you research it.

Some of the most pleasant hours we two authors of this book have spent with metal detectors have been related to searching for battlefield relics. In the summer of 1877 the Nez Perce Indians undertook their historic trek to Canada. Fierce battles between them and Army troops took place at many locations. Over the years we have searched many of these sites with various kinds of detectors. It is truly amazing how much more effective today's modern detectors are than those with which we were so well satisfied just a few years ago.

The soil at most of these battle sites has a high content of mineralization, and the terrain is generally rocky and uneven. The first challenge for a detector is to achieve precise ground balance that permits faint signals to be heard rather than background chatter. Secondly, searchcoils must be capable of operation at various heights because of rocks and other obstructions.

A third problem we encountered at the Nez Perce locations concerned hot rocks, those geological "freaks" that cause even the finest modern detectors to signal metal falsely. Since modern detectors with discrimination enabled us to deal with these little pests, we suggest that you employ such a detector for battlefield searching. Additional problems may come with ground balancing and excessive operating heights, but your modern instrument can overcome this when handled properly. Plus, it is always good to have discrimination . . . especially when you need it.

Research

How to locate areas where you can search for relics? Your answer, again, is research . . . research and more research. Oftentimes, all the research in the world cannot answer the question of whether a battle might have occurred in a particular area or whether gunfire took place at that precise location. Only your detector can prove the locations of battle or gunfire by locating bullets, cartridges or spent projectiles.

Research is particularly important since so many of the "obvious" locations to search for relics are now located in various state and national monuments and parks where the use of metal detectors is either banned entirely or highly restricted. You'll need to develop locations about which only you and few others know . . . perhaps you can develop these by learning more about the history of your ancestors.

Indeed, metal detectors can help you rewrite history. This was proved by the extensive survey taken at the Custer battle site at the Little Big Horn River in Montana. Study of the relics found by metal detectors helped increase the knowledge of where troopers and Indians died, but, more importantly, also changed perceptions of how they died. Evidence proved that Army weapons captured by Indians during the battle were turned on Custer's Cavalrymen in the final frenzy of the massacre.

Unlike the cache hunter, who is searching primarily for monetary reward, there are other reasons to search for relics and battlefield souvenirs. Some relic hunters are always looking for evidence to prove history, while others seek significant objects to add to a personal collection. Of course, many relics are sold, some for surprisingly large sums, and most relic hunters search for all of the above reasons.

It is fascinating to read tales of the early day settlers of this nation who simply picked up relics while plowing cropland or discovered them under brush in old battlefields. Those days have passed long ago. Except for those washed up or uncovered by storms, most of the visible relics have probably been found. Those remaining are below the earth's surface . . . sometimes far below . . . and can be located only by the modern ground-balanced metal detector.

What Size Searchcoil?

What size searchcoil should you use? You may be thinking that if you seek a small bullet, you shouldn't use the large 12-inch diameter searchcoils. Wrong! Large, high quality searchcoils will detect almost any tiny projectile, button or coin that you might find in an old battlefield. Because you will need all the depth possible when searching for relics, we urge you to use the larger searchcoils. If, even with a new, modern detector, you find this large coil too heavy for a full day's scanning, use an arm rest. This accessory provides counter balance for the detector and enables you to use it with even less muscle fatigue.

Because so many battle sites are in low-lying or swampy areas, it is well to make certain that your searchcoil is submersible. Don't be confused by such designations as "splashproof" or "waterproof." You will want a searchcoil that can be completely submerged two feet or so . . . to the cable connector that attaches to the control housing. If in doubt about your searchcoil's capabilities, ask your dealer or manufacturer. All Garrett Crossfire searchcoils, of course, are submersible. Be certain to check with other manufacturers about submersibility of their coils.

We recommend that you operate your detector in its All Metal mode to make certain you are getting the greatest depth possible from your instrument. Of course, large amounts of trash in the ground may cause you to think again. If you're really not concerned about losing just an occasional target, search while using some discrimination—but, with only a slight amount dialed by your controls. You'll miss small iron objects (including trash!), but will detect other targets such as those made of lead, brass and bronze. Of course, even using discrimination you will find coins to great depths with modern detectors.

Now, most professional relic hunters would be aghast at reading that last paragraph. They use absolutely NO discrimination. Here's one reason why: If a valuable coin or small military insignia is lying right next to an iron object and you are using discrimination, the iron object may cancel out the coin, causing you to miss it.

Ground Balance

In relic hunting and battlefield searching, it is of the utmost importance that you ground balance your detector *precisely*. As you have learned, this is an easy matter with modern instruments. If your detector offers automatic ground balance, go ahead and try it out first. Then, test your results by lowering the searchcoil to operating height. If you notice an excessive amount of audio change because of ground mineralization, use manual controls to ground balance the instrument as precisely as you wish. When you do this, you'll know that signals you receive from your detector come from targets and not from ground mineralization.

When they are properly ground balanced to compensate for minerals present in the soil, modern metal detectors can be operated with the searchcoil scanning at a height of up to four or more inches. The depth capabilities of these instruments permits the searchcoil to be held this high to clear weeds and brush as well as such objects as stumps and rocks. Of course, the searchcoil should always be held as close to the ground (and the targets below it) as possible for best results.

You perceptive treasure hunters already know what comes next, but I'll repeat, anyway: For maximum depth and sensitivity, use headphones and set your audio controls for the faintest threshold you can hear. This advice has been proven worthwhile over the years.

As, has this advice: *Scan slowly* to cover an area methodically, completely and thoroughly. Grid search an area if you can by tying lengths of ropes in grid squares and carefully searching each square before moving on to the next.

Another seemingly obvious tip is to double-check your hole – always! Just because you've dug up one relic doesn't mean that there can't be another in the exact spot! Often, battlefield relics – especially bullets – are discovered in "clusters." They were probably dropped or lost that way.

Of course, you should always double check every hole no matter what kind of targets you are hunting.   If, for example, you're seeking large objects with the Depth Multiplier attachment, always check your holes with a smaller searchcoil. Or, you can use a hand scanner such as Garrett's Little Treasure Hunter to make certain that smaller valuables aren't hiding in them.

Even with the modern metal detectors available today, the most important aspect that separates success and failure for the relic hunter is the skill of the hobbyist . . . the relic hunter's ability to operate his instrument to its maximum potential. It is very important that every relic hunter develop his or her skills to the maximum as they pursue relic hunting. They must also understand everything about the detectors they are using, including the potential of the instruments.

Just as we advised you to find battlefield sites through research, research, research . . . we now urge that you practice, practice, practice to understand all that there is to know about your metal detector and the way it reports its findings. Read your Owner's Manual so often that you can almost commit it to memory. Write or call the manufacturer if you have additional questions. Learn what to expect from your metal detector so that you will know what it is telling you in the field. Remember that your metal detector will always signal you about exactly what it believes to be lying beneath its searchcoil. These signals, of course, are based on the detector's circuitry but are regulated by ground balance and/or discrimination circuitry controls built-in by its manufacturer and adjusted by you. Your metal detector will never lie to you, but you must interpret the truths that it reveals.

Be Careful!

Because any battlefield might contain explosives, take all necessary precautions. Any time you dig into an object of any size, treat it as an explosive until you know otherwise. If in doubt, consult an authority . . . quickly! Remember that even old cannonballs, shells and mines that have been in the ground many years can still explode. Old guns, too, can still fire . . . especially if they are loaded! Don't get into arguments with explosives of any kind. It might prove dangerous to your health!

The same need for precaution holds true for underground power cables. If you're detecting to great depths with the Bloodhound or with any other large searchcoil, you might occasionally find an underground cable. Consider that it is carrying high voltage electricity until you are informed otherwise. Stop digging immediately! Contact the appropriate authorities, and inform them fully so that they can inspect your site and cover it properly.

Of course, you'll want to cover all holes properly, no matter where you hunt. This is true whether you're on prominent display with your detector and digging tools or whether you're searching in a secret location only you know about. In public you're the example of our hobby, and we always want our "best foot forward." Out in the boondocks you don't want to leave any evidence that could lead other relic hunters to your secret spot. You know you haven't found everything that is there!

Chapter 9
Antique Bottles

Old bottles are literally everywhere – and they can be found by the millions! Except for occasional eye appeal or other fascination, most of these are virtually worthless to anyone. There are those collectors who prize bottles, however. For them, there are some bottles that are truly priceless. All treasure hunters are urged to be aware of the potential value of old bottles. Do not overlook them under any circumstances.

Be sure to acquaint yourself adequately with this widespread hobby of bottle collecting. Believe us when we tell you that as a treasure hunter, you will certainly be afforded the opportunity to collect them . . . time and time again. When you hunt off the beaten path with a metal detector, you may not always find coins, jewelry, gold nuggets or what other targets you may be seeking. But, you will find bottles. Yes, you can depend upon always finding bottles. Never disparage these discoveries. They can result in a beautiful collection for you and your family, and they can also result in valuable monetary rewards.

The search for antique bottles – collecting them and then presenting these artistic objects in colorful and creative displays – long predates the invention of even the earliest metal detector. Collecting bottles is a hobby that stretches back into antiquity. Yet, bottles remain an attractive target sought avidly today by many treasure hunters.

The hobby of bottle collecting goes hand in hand with that of metal detecting. Valuable antique bottles can particularly be found in conjunction with relics in ghost towns and at old dump grounds. When searching for objects with a metal detector, do not overlook valuable antique bottles . . . whether you prize them for their appearance or for a monetary value which can sometimes prove surprising.

How can you tell a valuable bottle? Scan the various treasure, collectible and diver magazines occasionally for displays of bottles and their values. Learn to look for the different and the unusual. Have some idea of what bottles are valuable and which are just "pretty."

But, why not bring home those bottles you find that are just "pretty?" You'll develop a beautiful collection for yourself, and you'll probably find that you've retained some that are quite valuable.

Why does man search for bottles? Why has he been doing so since the very dawn of time? First, it was strictly for utilitarian reasons. Bottles were crude and clumsy containers, scarcely utilitarian and certainly not worth taking a second look at for any aesthetic reason. They were seldom discarded because, no matter what their age, bottles could be quickly put to use for transporting and storing liquids. Then, as beauty became a more integral part of the glassmaker's art, bottles themselves evolved into objects to admire and to value. Even though value as measured by cash transactions rises and falls like that of all collectibles, continually improving prices have been paid over the years for bottles . . . especially those that are old, oddly made or attractive or unusual in some other way.

Facing
Charles Garrett searches for relics at a 17th century fortification on a Caribbean island. Above are shown typical battlefield discoveries.

Over
These weaponry relics were found in the attic of an old house. Inscribed on the powderhorn is a sketch of the famed Alamo, where Texas patriots died.

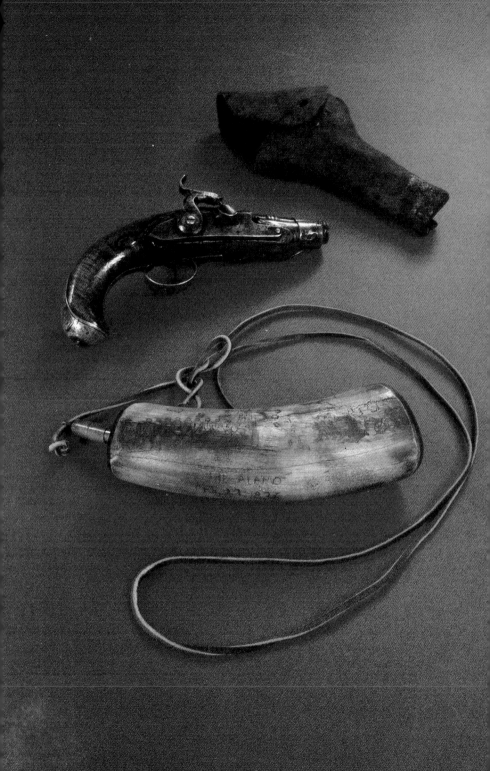

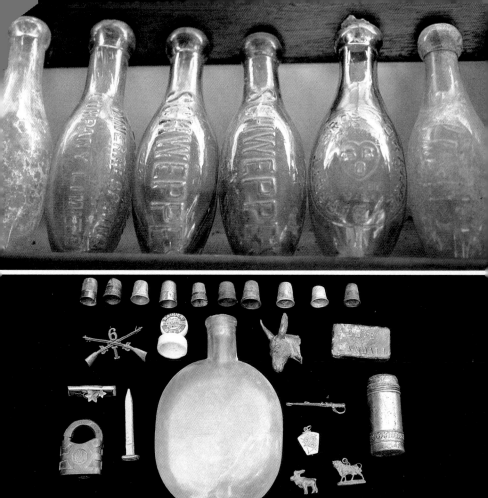

Bottle-Making is Ancient

The practice of bottle-making can be traced back thousands of years. Early glass manufacture was slow, costly and required hard work. Tiny jugs, jars and cosmetic cases were crudely fashioned and were prized by the ruling classes. Those who could afford these early glass containers considered them as precious as jewels.

Merchants soon realized that such liquid commodities as wines, honey and oils could be stored and carried more easily in glass containers than in wood or clay. It was not until about the time of Christ, however, that man learned to blow glass. The resulting bottles were the first of those carefully shaped, beautifully colored, hand blown objects that are so prized today by collectors.

In the twenty centuries that have since passed, literally billions upon billions of bottles have been manufactured, first by hand and later by machine. These vessels were initially designed to package primarily food and beverages, but in later centuries they became just as important to transport and present for sale medicines, cosmetics and chemicals. In earlier days, reuse of bottles was quite common. But, even then, many were discarded upon being emptied . . . others were discarded, either accidentally or intentionally, while still partially filled.

Facing
The **right** kind of stream is ideal for electronic prospecting since water generally deposits gold on bedrock along the bottom.

Over
Bottles such as these shown here can provide beautiful prizes to treasure hunters with metal detectors. Always watch for unusual or especially lovely bottles.

Few of today's bottles are reused; once emptied, they are usually either saved or quickly discarded. Of this great multitude of bottles that are thrown away daily, most are processed immediately by commercial waste disposal plants and destroyed. Others, however, will somehow find their way to join those of months, years and centuries gone by and lie by roadsides, in streambeds, in dumps and wherever mankind has left its debris. Given the beauty of many bottles and the unique nature of others, is it any wonder that collectors have come to admire and treasure them?

Obviously, then, there are literally limitless numbers of bottles to be found. The key to successful bottle hunting and collecting lies in research which is so important to all phases of treasure hunting. Good, productive locations must be found, and more of these can be located than can be searched effectively in the lifetime of any bottle hunter.

Bottles can be found today where people once lived or congregated. With just a little study of history these productive sites can be located. Where were the towns, military posts and forts, waterfronts, bridges, fords and mills once located? Where were the sites of yesterday's commerce and industry? Old newspapers, maps, historical accounts and other sources will answer all of these questions and more.

Plan to find many of your bottles associated with the water. Perhaps this is because bottles are less likely to become broken after they have been thrown into a stream or body of water. Of course, bottles protected by water are not exposed to the erosive effects of wind and air. So, always keep your eyes open for the possibility of valuable antique bottles when you are searching in a stream, lake or any other body of water.

Some bottles survive better in water than others. When bottles are lost in the water, they become covered with sand, entrapped in sediment or, in the ocean, encrusted with reef organisms. You'll find bottles that are badly corroded while others show only mild film residue. Some will become pitted while others develop a heavy encrustation.

The laws of physics often provide real surprises for bottle hunters. Imagine, looking into a stream and discovering lying atop its bed of sand a bottle that was tossed there 100 years

110

ago? How did this happen? You know that in a century a coin would have sunk into the earth, perhaps even below your detector's capability to detect it? How did this bottle remain on the surface? Or, has it just been somehow dug up and tossed to the surface?

Specific Gravity

Specific gravity and size and shape have a lot to do with the depth at which bottles can be found. The specific gravity or weight of any object compared with that of its surrounding materials governs the pull of the earth's gravitational field on that object. Because a coin is heavier than the earth surrounding it, the coin tends to sink at every opportunity. This occurs even faster in water, where wave and wind motions tend to loosen bottom materials, allowing heavier objects to sink.

Since bottles are made of various earth materials such as sand and clay, their specific gravity can be the same as the river-bottom sand and mud. Gravitational pull upon such manmade objects is the same as that upon surrounding materials. Thus, bottles often do not sink but truly "float" on a sea of similar sand, clay and gravel.

The result is a 100-year-old bottle appearing beneath you on the sandy bed of a stream.

Lakes can be treasure storehouses of bottles – of all shapes and sizes. Just because you cannot see them at first glance does not mean they are not there and have not worked their way out into deeper water. They will sink into mud and silt where you can recover them with just a little digging. Be careful, however, not to expose bare hands to fishhooks, lures, wire and other sharp objects that will also be buried in the silt.

Near the oceans where rivers rise and fall with ocean tides you can have your best success in finding bottles during times of low tides. Watch for bottle shards, broken china and other debris. In shallow waterways work around exposed boulders and sunken logs where bottles collect. Work fords, ferryboat sites and old swimming holes and under and around bridge sites. Also work downstream from these locations. At bends of rivers where the flowing water slows moving objects bottles will tend to fill and sink. Look for areas where embankments are washing away. Particularly in populated areas, these embankments that erode will expose old bottles.

You may not have thought of collecting bottles as a likely hobby for the metal detectorist. There's a fascination and charm about searching for and finding them, however, that exists in no other field of treasure hunting. They make handsome and charming displays. Their literally limitless variety of shapes, colors and designs enable collectors to fill shelf after shelf with striking arrays.

Finally, there is real profit to be found in bottle collecting. Many bottles will command decent prices at all times, with the higher valued specimens increasing dramatically in value as the years pass.

Be always on the alert for the unusual bottle, for the bottle that is different . . . that attracts your attention because of its nature or particular beauty. You may have discovered a real treasure!

Chapter 10
Electronic
Prospecting

The single word, *gold,* has had a magical effect on mankind since the very dawn of creation. Whether whispered, shouted or offered in awe, the very name of this precious element can instantly create admiration, avarice, satisfaction, joy and a host of other emotions. Gold was one of the first known of all the metals, and it has long been valued among the most precious.

Prospecting for gold was one of the earliest uses of commercial metal detectors. Military veterans were quick to see how the electronic devices they had encountered in the service could now help them find gold. They expected detectors to discover new locations for mining and panning while pointing out nuggets and specimens that had been overlooked by early-day prospectors.

Today's modern electronic metal detector represents an even greater advancement over those early instruments than these first commercial models represented over the eyesight and instinct of the old-time prospectors. Vast numbers of articles and books have been written concerning electronic prospecting – including several by us, collectively and individually. Much of their material, however, deals with detectors that have been made obsolete by the new instruments. This chapter, on the other hand, discusses only modern instruments. It will seek to summarize the basic fundamentals of looking for precious metals with today's metal detectors, including computerized models. After applying these principles, many of you will probably want to learn more about this phase of treasure hunting by reading our prospecting books and those of others.

To say that there can be no greater satisfaction than the discovery of gold may be gross. Whisper the word to yourself, however; say it aloud. The sound is beautiful. When your metal detector alerts you to the presence of gold, you will hear music of the angels indeed!

Three Basic Rules

Observe these basic rules for prospecting with a metal detector. They will certainly not guarantee you success, but if you learn and apply them, your chances of finding gold will be enhanced. Certainly, if you are persistent, you can locate and identify any precious metals you seek:

1. Select the proper type of detector.
2. Always be patient.
3. Look where gold is known to exist.

In discussing the "proper type" of detector, we refer, of course, to the operating characteristics of the detector rather than the brand name. There are several excellent nugget detectors on the market, but we, obviously, recommend the new Garrett A3B-United States Gold Hunter, with its legendary Groundhog circuit that has found tens of thousands of ounces of gold in many countries and around the world, or the new Grand Master Hunter with its microprocessor-controlled, computerized circuitry preprogrammed at the factory. These two detectors are capable of locating even pin head-size nuggets and will identify hot rocks properly.

Patience takes many forms in treasure hunting. You should understand your detector and its capabilities fully. You must also learn its limitations to become proficent in its use. Read this book carefully along with the Owner's Manual supplied with your detector. After you have practiced with your detector, read both books again and study them. Never get in a hurry. Remember that gold is always discovered during your *next* searchcoil sweep. But, you must know *where* to make that next scan!

It is impossible to find gold or any other precious metal where it simply *does not exist*. Prospectors who have found gold with metal detectors have demonstrated wisdom and patience, plus a great deal of *research*. Always stick to known, productive mining areas until you are very familiar with electronic prospecting, generally, as well as your detector's operation. You must be able to understand how it signals the presence of mineral zones.

To achieve the greatest success in finding precious metals, the ideal metal detector for prospecting is a sensitive, deep-seeking instrument that has been designed for prospecting and proved in field use. It must offer precise (adjustable in the field) ground balance and properly calibrated discrimination circuitry. Such a detector must be calibrated correctly by the manufacturer for metal vs. mineral determination and hot rock identification. With power to penetrate the toughest iron mineralization, it should have the availability of a complete range of searchcoils from about 4 1/2 inches in diameter to about 12 inches.

Garrett's A3B Gold Hunter and Master Hunter 5 both offer 10-turn controls for both ground balance and audio threshold control. These permit the extremely precise adjustment demanded for finding tiny nuggets in heavily mineralized areas. Similar ground balancing and detection characteristics are offered on Garrett's computerized Grand Master Hunter, which utilizes push buttons for precise control of ground balance, so vital in prospecting.

It is possible, to some extent, to use *any* ground balancing detector to find precious metals. Most such modern detectors can be operated even by a knowledgeable beginner in areas where the older models would have proved almost useless in the hands of an expert.

Automated – or, "motion" detectors, as they are sometimes called – can be used for nugget hunting. When using such an instrument, however, you must resign yourself to overlooking tiny nuggets. And, many of the motion detector's discrimination controls are not properly set at the factory for metal/mineral discrimination. Even when the discrimination controls are set at zero (all metal), the detector usually still provides considerable discrimination that might cause some gold to be lost.

Depending on existing mineral conditions, you might achieve some success, but we can almost guarantee that you will overlook nuggets, ore veins and the like that would be found with the *right type* of detector. The instrument with properly calibrated ground balance and discrimination that has been proved by successful testing and use in the field will offer superior performance in detecting small and large nuggets and both conductive and non-conductive ore (predominantly magnetic) veins.

Detectors that discriminate or reject targets by operator manipulation, such as "whipping" or operating at a prescribed speed, are virtually useless in prospecting. You should also remember that you will often be operating in areas (mine tunnels, for example) that are too cramped to permit whipping.

During the early days of metal detecting, much literature was produced (by these authors, too!) on prospecting with detectors. The majority of that writing revolved around the ancient, but trusty, workhorse of the industry, the beat frequency oscillator (BFO) metal detector. Both of us can become almost romantic in recounting our successes with it. We loved that old detector, but the simple fact is that the BFO is obsolete today. It is no longer manufactured. Anyone who has been successful in the gold fields with a BFO will be able to magnify this success MANY times over through use of a proper modern detector. Our experiences conclusively prove the truth of that statement.

Prospecting Techniques

Every metal detector **should** be calibrated and tested before it leaves the factory. (All of Garrett's detectors are calibrated and tested.) With such a calibrated detector you can check ore samples accurately in the Discriminate mode simply by setting your discrimination control to its lowest setting.

Before you take your detector into the field, experiment with it on the bench to learn how it will react to various ore samples. Use metal and mineral samples you have made or purchased. Get a very high grade specimen of each. By conducting your own bench analysis you can become familiar with the type and amount of detector response to the various materials.

If the ore you are checking has a predominance of metal in detectable form (such as a solid silver dime, for example), you will hear a sound increase as the sample is moved across the searchcoil. If your detector is equipped with a sensitivity meter, you will see a positive pointer movement indicating the presence of metal. On the other hand, if the sound has a tendency to "die" slightly, your sample contains a predominance of mineral or natural magnetic iron (Fe_3O_4). As mentioned previously, this does not mean that the sample contains *no* metal, only that it contains *more* mineral than it does metal. If the sample contains *neither* metal nor mineral, or if it has electrically *equal* amounts of both, you will receive *no* response on your detector.

Always remember that metal detectors do NOT respond to many types of ores. Only those containing metal in conductive form in sufficient quantity to disturb the electromagnetic field will cause the detector to respond positively. Continually practice bench testing to become more familiar with the type and amount of response your particular detector gives to high grade ores. A little time experimenting can save you many hours of confusion when you are in the field – and it may keep you from missing valuable nuggets, ore samples or veins!

Hot Rocks

This term, "hot rock," is a great bugaboo for the beginning electronic prospector – and it's no joy for the veteran, either! Some detector manufacturers pitch entire advertising programs (accompanied by extravagant claims) on the abilities of their instruments to deal with these little pests. Many books on electronic prospecting make hot rocks a major issue.

They are given undue importance! Believe us. You may never encounter one! Not likely, however. To use a cliche and to use it aptly, hot rocks are the absolute bane of a prospector's existence. There is really no getting away from them if you are diligent and persistent in your search for gold. Furthermore, they can drive you crazy if you let them. Let us emphasize how vital it is that you learn to understand these little pests and to know how to respond to them properly. Have confidence in your ability and that of your detector to deal with them. Understand, first of all, that identification of hot rocks and hot spots is no problem whatsoever to a modern, properly calibrated detector. Then, meet your hot rocks with confidence, not with fear!

A hot rock is a geologic freak. It is a rock that is located out-of-place and does not correspond to the metal/mineral composition of ground and the rocks around it. The hot rock, therefore, causes a problem with your detector which you have

117

carefully ground balanced to existing ground mineral conditions. Your "hot rock" contains a quantity and/or density of non-conductive iron mineral that responds as METAL, usually because the ground balance controls of your VLF detector are not adjusted for that particular rock.

Hot rocks can be found in jumbled rock piles as well as in ground areas that look as if they have never been disturbed. They contain no valuable conductive material. Their iron mineral content, however, is so exceedingly different from that of surrounding rocks that they disturb the delicate ground balance of your detector and present themselves in its electromagneric field as metal targets.

When you are operating with your detector in the All Metal mode, the signal you receive from a hot rock will be the same as you would receive from a metallic object. The signal is positive and unmistakably metal. Of course, you should first try to identify any target before digging.

How to Identify Hot Rocks

Let us make clear again that identification of hot rocks and hot spots is no problem whatsoever to a modern, properly calibrated detector. The procedure for identifying mineral hot rocks (and hot spots) is a simple one, but it will require practice from you. To check to see if the "metal" response is metal or mineral, pinpoint your target with the detector in the All Metal mode. Then, move the searchcoil to one side, lower it slightly or set it on the ground and switch to the Discriminate mode of operation. Your discrimination controls should be set to zero or to the level specified by the manufacturer as the calibrated level for ore sampling. Audio retune the detector, if necessary, according to manufacturer's instructions. (Garrett detectors retune automatically.) Now, with a constant sound (threshold) coming from your detector, pass the searchcoil back over the target. Keep the searchcoil at the same distance from the ground, as just discussed. Maintaining constant searchcoil height may be difficult at first, but you can accomplish it with practice. If the sound level decreases (or goes silent), your target is magnetic iron ore or oxides. These are the *only* substances that will cause the signal to stop. When this happens, ignore the target, switch back into All Metal and continue searching.

You have just exposed a hot spot!

If, on the other hand, your signal increases or remains steady, the target should be investigated. Increase your variable discrimination control (usually by turning the knob clockwise) to determine the amount of conductivity in this target. This procedure makes it obvious why you should not attempt to prospect with a detector that has a fixed or programmed discrimination control. If you have previously practiced with varying discrimination, you already know the approximate point on the control where worthless pyrite and low grade (conductive) ore will be rejected. If you continue to receive a positive response after you have passed this setting, it is very possible that you have discovered a non-ferrous pocket, a vein of conductive ore or a nugget.

You may have just struck it rich!

Hot rocks are "freaks" or "oddballs" of nature. They should not be where they are, and they should not cause your detector to react the way that it does. But, they do! Simply stated, however, hot rocks are "there," and we must deal with them. With a properly calibrated detector that properly discriminates you can identify them quickly, and they will become to you exactly what they are to veteran prospectors – tolerable pests.

Nugget Hunting

Nugget hunting is the most popular form of electronic prospecting and offers immediate rewards when you are searching the proper areas. Look for nuggets where they have previously been found. That is the key!

If you are a good coin hunter, you already know how to search for nuggets. Use the same techniques.˙ Only use them in gold country instead of in the park or schoolyard. We recommend the All Metal mode. Most professionals do not use Discriminate mode because even small amounts of discrimination can result in losing nuggets. They know, too, that the target identification meter on the best modern detectors will help them to identify *all* detected targets, provided that the target is both large enough and close enough.

We recommend that you use the a searchcoil with a diameter

of eight inches or so or a smaller coil, such as the Garrett Super Sniper, when searching for nuggets. Of course, the large (12-inch) searchcoil can be used because it will penetrate deeper, yet still find small nuggets. Whether you use it or not depends on the area and upon the types of nuggets that have been found there. Larger nuggets at extreme depths, for instance, may require the large searchcoil.

Headphones are an absolute must in nugget hunting because the small size of many nuggets results in only faint audio signals.

The most troublesome aspect of searching for nuggets – depending upon your search area, of course – is the detection of junk and mineralized pebbles (known as "hot rocks"). We have already discussed how to deal with them. Concerning other junk . . . since we do not recommend the use of discrimination while searching, you should learn other techniques for determining if you have detected a possible nugget or just a piece of iron.

First of all, be prepared to dig some junk. You'll be working, most likely, in or near old mining camps where countless pieces of junk iron were discarded. Unfortunately, all the many boot nails that may have been lost here cause your detector to respond all too often with a sound just like a good target. Luckily, boot nails are usually shallow, and you can spot them rather easily once you learn how.

Another way to determine quickly if you've detected an iron object is to carry along a strong magnet. At your point of detection, rub this magnet in the dirt. Lift up the magnet and your target will be stuck to it if it is, indeed, iron. Several good digging tools and rakes are available that contain a strong magnet in the end of the handle for quick recovery of iron objects.

Be sure to carry a pouch with you to carry out all detected objects. Other hobbyists will follow you, and there is no need for them to encounter your trash. Who knows? You may search the area again, and you certainly don't want to have to bother with the same junk targets.

When you are using your modern detector for electronic prospecting, use the same techniques you learned when coin hunting. Adjust your audio for faint sound. Use a good set of high quality headphones. Scan slowly and methodically to cover all areas of the ground. Be certain to move large boulders aside so that you can scan under them. Scan up against brush and other obstacles. If you find the ground extremely uneven and cannot maintain a constant searchcoil height, you may encounter problems with ground balance and ground mineralization. If so, simply elevate your searchcoil a few inches, set your manual ground balance controls at that height and continue scanning. You will not lose depth. In fact, because of the better quality of your audio, you may even detect nuggets more deeply and more accurately.

If the area you are searching is so littered with iron junk or hot rocks that it is really troublesome, experiment with small amounts of discrimination. Dig quite a few targets you detect at zero discrimination and study what you have found. Then use the discrimination techniques discussed earlier in this book, but advance your controls no farther clockwise than necessary to reject the most troublesome of your pests. True, you may lose a nugget. You may decide to chance this rather than dig hundreds and hundreds of worthless nails or hot rocks. Who knows . . . the time that you save by not digging junk may let you search that extra bit of ground where a real beauty of a nugget awaits your modern metal detector.

We suggest you manually adjust your detector for the most precise ground balance. If you like, try automatic ground balancing first. If, when scanning, the audio changes by an amount that you feel is too great, manually ground balance the instrument. Precise adjustment of ground balance will enable you to deal with ground mineralization more properly.

There is no magic to locating nuggets! Finding them just requires patient application of all the skills you have learned with your modern metal detector. Nugget hunters are successful because they are willing to work hard and hunt for long hours. Try it. It will pay you dividends!

Searching Mine Dumps/Tailings

You can locate mine dumps in all shapes and and sizes and in varying conditions. Each one probably contains SOME material of value. Essentially, a mine dump is nothing more than a storage, waste or holding area for rocks that have been removed from a mine. They can be small piles or huge mountains of rock.

Most mine dumps were meant to contain waste, the rock that surrounded ore material that went to the crusher. Many times, however, miners unknowingly discarded precious ore that is conductive. The use of a metal detector offers the only practical means of discovering that ore. The naked eye is useless for such a task, but the probing eye of a modern metal detector can look within rocks to find those that are valuable.

How do you search mine dumps with a metal detector? You can either conduct general scanning over a large area of rock surface or select a specific spot and test likely looking rock specimens from that area.

When scanning mine waste, remember what lies beneath you – a jumbled mass of rocks that have been removed from their natural geological locations and thrown together. Perhaps each rock contains a different amount of iron mineral. Because your detector will analyze all conductive matter beneath its searchcoil at any time, it may have difficulty with such non-uniform minerals. Your challenge will lie in proper ground balancing, but there are tricks we recommend.

Precise ground balance is essential. If your detector offers automatic ground balancing, you should try it first and scan the immediate area of rocks in which you are interested. If your audio changes significantly, try to manually adjust the detector more precisely. If that fails, switch to a larger searchcoil and/or scan with the searchcoil held a greater distance away from the rocks. Try all of these techniques, even lowering the volume of your audio to silent to reduce the chattering effect of this mass of uneven minerals.

Use your All Metal mode and scan as you normally would for metal objects. Wear headphones and scan methodically. Be thorough in analyzing each signal from your detector. It could be a manmade object or a piece of gold or silver ore. Be careful not to lose your specimen, especially in a pile of loose rock. When digging for the object that caused a signal, you could dislodge it to fall farther down into the pile. Your recovery then becomes more difficult, if not impossible.

Use a plastic gold pan. When you get a signal, rake the top layers of rock into your pan, then scan the pan to see if you have recovered that object causing the signal. If it is still in the pile, discard those rocks in the pan and keep trying until you know exactly where it is. Using the gold pan, you will stand a better chance of not losing your target and probably recover it more quickly. More on this subject can be found in Chapter 12.

When scanning in the All Metal mode, listen for even the faintest signal through your headphones. This is important because it takes only a little gold to make a valuable ore specimen. After you have searched an area and received no signals, use your shovel to remove a layer of rocks and continue searching. By removing this top layer you are placing your detector's search-coil closer to possible paydirt.

No one specific location – the top, sides or lower levels – in a mine dump can be said to be consistently better as a place to recover nuggets and/or ore. As you gain experience and train yourself to analyze rock and rock structures, you will be able to pick out the best locations more easily. Remember that the early day miners did little more than follow a vein into a mountain. They piled the associated rock in a cart and dumped it. Careful study will indicate the type rocks that require scanning. When you find some of these rocks together, lay your detector flat and operate in its Discriminate mode. Make certain discrimination control(s) are at their lowest (usually, to the left) settings. Pick up individual rocks and bring each quickly across the searchcoil's bottom. If any rock causes a signal increase, keep that specimen for further study.

Always use caution in digging and climbing, especially on steep piles. You could easily create a rock slide and injure yourself and others.

Dredge Pile Searching

In their earlier efforts to recover precious metal, prospectors came into gold country with huge earth-moving devices and literally scooped out streams down to bedrock. Countless tons of rock were removed in a relentless search for gold. Since gold is one of nature's heavier elements, the actions of wind and water work with gravity to pull it down to the lowest possible level. This is especially true in stream beds where the gold finally reaches bedrock. There it may lie for eons, awaiting the prospector . . . whether he has a huge dredge . . . a simple pan . . . or a sophisticated modern metal detector.

Prospectors with dredges scooped up vast quantities of rocks which they immediately began processing, first by classifying according to size. Huge conveyor belts transported waste material and large rocks to locations on the banks of the stream where they lie today, often in unsightly heaps. These are the dredge piles that have been successfully searched by so many electronic prospectors. Many of these piles contain more than just waste. Occasionally, large nuggets escaped the eyes of the dredgers and their sorting equipment and went into the dredge piles. Small nuggets passed through in mud or clay balls. They are still there awaiting your detector.

Scan these piles as you learned to scan mine dumps. Ground balance your detector as precisely as possible and listen closely for increases in sound. From time to time, dig down into the pile to expose your searchcoil to a different layer of material. Searching individual samples with the ore test method is not effective here since you will not be able to find any accumulations from one specific area in the earth. Your dredge piles are truly just piles of individual rocks that were tumbled from many different locations . . . first by nature and later by the dredgers.

Facing
The authors admire a nugget in dredge piles alongside a creek bank found with the techniques of detector and gold pan explained in this book.

Over
Roy Lagal searches with a small Super Sniper searchcoil to look between rocks beside a stream for gold nuggets deposited here by high water.

124

Grid Patterns

Searching in grid patterns is a method that has proved successful in field prospecting when looking for veins. It offers the opportunity of approaching a vein or other type deposit from two different directions, avoiding the distinct possibility of walking parallel to a vein but never actually crossing it. Set up a fixed grid or crisscross pattern and sweep your searchcoil across it in wide, even strokes.

Always adjust your detector in the All Metal mode, balanced to the mineralization of the surrounding ground. Use manual audio tuning or the factory-set ultra slow auto audio retuning mode. Walk slowly in as straight a line as topography permits, scanning a wide path with the searchcoil ahead of you. When you reach the end of one of your "lines," turn and walk a parallel path in the opposite direction approximately two feet from the first path. Continue until you have covered the area selected. Then, repeat this procedure, except walk parallel paths at 90-degree angles to the first set of paths. Now, you have completed your fixed grid or crisscross pattern search of the area.

The so-called "two box" or RF searchcoils are especially suited to this task because they are designed for searching

Facing
With the exposed quartz vein behind indicating the possibility of gold, the authors carefully scan the surrounding area for it and other conductive metals.

Over
When miners excavated precious metals, they had only their eyesight as a prospecting tool. The modern metal detector can find ore that they overlooked.

deeper. Garrett manufactures the Depth Multiplier, which is also called the Bloodhound. This searchcoil is especially easy to use because it requires no intricate electronic adjustments. Just attach it to the Grand Master or other Garrett Master Hunter instrument and start searching. Another advantage of the Bloodhound is that it is not designed for detecting small objects. Therefore, you will spend far less time digging "junk."

When you are walking your grids and your detector speaker sound increases and/or the meter pointer shows any increase, notice the intensity and the duration of the increase. You may have discovered a vein or ore pocket beneath you.

On the other hand, your detector may have changed its tuning due to atmospheric conditions, interference, bumping of controls or some other reason. Do not touch your detector's controls. Rather, return to that point where you were scanning before you noticed an increase. If the speaker and/or meter decreases to the previous level, your detector's positive response was caused by conditions in the ground, not some detector or operator problem.

Continue to retrace your steps. As you reach the point where the sound changed earlier, it should again change if you are detecting an ore vein or pocket. As you continue walking, pay close attention to the detector's audio responses. They should either continue to increase or drop off to your initial tuning level. If responses are "increasing," your ore is getting RICHER. When the responses return to your initial tuning level, you have walked on over or past the ore deposit. Plot or map the deposits or veins. If there are several, note where they cross or crisscross each other beneath the surface.

Prospecting for gold is probably the most romantic use to which you will ever put your metal detector. The rewards can be truly unlimited. An amateur prospector in Australia, using a Garrett detector, discovered a gold nugget that later sold for a reported one million dollars. Even if your nuggets are tiny – or, if you only get a "touch of color" in a pan – total or eventual rewards for electronic prospecting can be vast. Enjoying the great outdoors in beautiful gold country is a magnificent experience. When you can combine this with the prospect of sudden – and, perhaps, great – wealth, your rewards are vast indeed. The electronic prospector may truly be the most blessed of all metal detector hobbyists!

Chapter 11
Rocks, Gems and Minerals

The modern metal detector truly adds another dimension to the capabilities of the rockhound, that familiar veteran of the outdoors who scrambles up and down the sides of mountains with a hammer in one hand and a sample bag in the other. In addition to that invaluable rock hammer and ever-questing eyes, the rockhound now has an instrument that provides a vital "look" into the very heart of rocks.

Early metal detectors with their limited capabilities and complicated operating methods often baffled many rockhounds. They did not have time to develop the expertise required to understand the operation of these detectors and the confusing signals they often produced. Rockhounds could be certain the detector was telling them "something;" just "what" it was saying was an entirely different matter. Believe us, detectors have changed! If we could emphasize one point, it would be that today's modern detectors – though incredibly more capable than the earlier models – are also far *easier* to understand and to operate. The rockhound who, for one reason or another, has never considered the metal detector as a primary tool would do well to reconsider.

The knowledgeable rockhound now uses a metal detector as an aid, not as a complete searching tool. The wily veteran, wise in the ways of metals and minerals as they present themselves in rocks, understands that the modern detector offers no magic wand and treats the instrument as just another implement in a varied bag of field equipment. Yet, any rocks that appear of interest are invariably tested with the detector. This kind of testing and the resulting examination not only increases knowledge but also never fails to produce valuable specimens.

Of course, nothing can take the place of that valuable experience a rockhound has gained through years of identifying semi-precious stones and gems. The metal detector merely aids in the location of conductive metal specimens that the human eye simply finds impossible to distinguish or identify. No longer must a rockhound risk overlooking a high grade specimen simply because of lack of instant visual recognition. The modern metal detector can easily look into the rock itself to help in its proper identification.

The simplest definition of metal to a detectorist or rockhound is "any metallic substance of a conductive nature in sufficient quantity to disturb the electromagnetic field of the searchcoil." Prime examples with high conductivity are gold, silver and copper. They and other non-ferrous conductive substances are considered metals. As an added tool for the rockhound, a modern metal detector clearly identifies those specimens that contain conductive metal in some form. For just a few minutes' work the rockhound, therefore, can wind up with a high grade metallic sample that has been passed over for years by others in the field.

Practice to familiarize yourself with the responses produced by metallic substances. Use specimens with which you are familiar, and you will be able to identify more quickly those samples which are unknown to you. Use the bench testing method that is described for ore sampling.

Modern metal detectors are ideal for identifying metallic ores. When searching in your All-Metal mode, and you get a positive reading, switch to the Discriminate mode (using absolutely no discrimination) and advance the discrimination control to determine whether the content is ferrous or a non-ferrous precious metal.

Your detector dealer may have a wide selection of ore samples and gems that are either well known or identified. You can also get such samples at rock shops. Practice to learn the abilities of your detector by experimenting with these samples that are conductive. Such knowledge will enable you to enjoy your hobby and increase your rewards from it. Do not minimize the possibilities for success with a good modern metal detector. There are vast numbers of hobbyists (and professionals) who use these instruments to search for specimens that have been missed by visual searching or that were unrecognizable to the naked eye. Some of these may contain precious metal that is worth a goodly sum at today's prices.

Ore dumps, rocks piles, dry creek beds, dredge tailings or any promising area may be searched with a metal detector. Always use the All Metal mode with the instrument precisely ground balanced to existing mineralization. Your instrument will then identify any specimen of predominantly metalllic content. Classify your specimen using discrimination controls in the Discriminate mode.

Always pay close attention to mine tailings and ore dumps when using your metal detector to search for gems. There could be high grade ore specimens, and the "extra eye" of your metal detector will call your attention to them.

If you are a rockhound, we urge you to use your modern metal detector. It can help you. To metal detector hobbyists we offer the invitation to become a rockhound and to search for semi-precious stones with your instrument. Our advice to both? Experiment! A whole new world will open up when you become familiar with the capabilities of your detector. What it can do will amaze you. You will soon be able to check promising rocks more carefully and precisely classify them according to conductivity. Those searched-out areas may prove to be not so barren after all!

We are constantly amazed at the lack of knowledge about metal detectors shown by so many so-called experts in the fields of minerals and rock-hunting. Time and again, we have witnessed specimens being sold at rock shows with the seller having no idea of the metallic content (and VALUE!) of the rocks being offered for sale. Occasionally we will seek to explain how their metallic content can be determined with a metal detector and are greeted with a complete lack of interest. "I don't believe in metal detectors," is a common response.

Three of us found several hundred pounds of silver ore one day in searching near Cobalt, Ontario. That evening, we met two rockhounds returning to their vehicle after a full day's search. In their pouches they had about ten pounds of ore. We asked them if they knew about metal detectors. Like most other rockhounds, they said that they didn't believe in them. We didn't have the heart to show them the specimens we had just found.

Well, becauses of successes such as this, we obviously believe in metal detectors! And, we have proved that they can be just as valuable to a rockhound as they can be to a coin hunter, electronic prospector or any other user of modern electronic instruments.

Chapter 12

Finding Gold With Pan and Detector

Agood, plastic gold pan is vitally important to the electronic prospector. In fact, today's gold-hunter, seeking the precious metal with a modern electronic detector, requires a gold pan just as much as the old 19th century prospector who had little more than a shovel and a burro in addition to his pan.

The modern metal detector and a good plastic gold pan have proved to be a *winning combination* indeed for many weekend prospectors. After your metal detector locates nugget deposits or promising panning locations, a good plastic pan is almost essential in their recovery. Yet the pan alone will not be satisfactory.

Believe us! Regardless of whether your primary intention is to recover gold by locating it with a metal detector, capturing it in a dredge or by sluicing, a gold pan will remain your *primary tool*. Pans will obviously be necessary in streams, but you will also find them useful far away from water, often in ways that a novice weekend prospector could never imagine. Dry panning is sometimes the only practical way to discover gold in desert areas. The pan may also be used for easy recovery of metallic targets signaled by your detector . . . objects that might be difficult or impossible to locate otherwise.

Simply stated, if you intend to seek gold in the field, it is vital that you not only have a good gold pan but that you understand its importance and know how to use it properly.

Just as important here is having the *right* kind of pan. Of course, any sturdy type of basin bowl can be satisfactory. Since the old prospector from the Western movies last rode his decrepit burro into a beautiful Hollywood sunset, however, pan designs have improved considerably over the old skillet he used. Today's gold pan is lighter in weight and offers greater speed in testing and classifying concentrates. It is also easier to handle and provides safer, surer results, especially for the beginner. Absolutely no experience is required for any individual to enjoy success in the first panning session.

During the gold rush days of the 19th and early 20th centuries, gold panning was more than hard work; it was back-breaking labor. Unless a panner was lucky, it was usually not especially profitable. Today, however, gold panning is much easier and far more lucrative. This has happened not only because of the increased price of gold but because of the modern gold pans that are now available.

The "Gravity Trap" Pan

Of course, we naturally believe the finest and most effective gold pan today to be the **"Gravity Trap"** pan. Invented by Roy Lagal (U.S. Patent #4,162,969) and manufactured by Charles Garrett's company, Garrett Electronics, its effectiveness has been proven by worldwide success and acceptance. Made of unbreakable polypropylene, the pan is far lighter and easier to handle than the old metal pans. More importantly, the Gravity Trap pan has built-in gold traps in the form of sharp 90-degree riffles. These riffles are designed to trap the heavier gold and allow fast panning off of unwanted sand, rocks and gravel.

The pan is forest green in color which has been proved in extensive laboratory and field tests to show gold, garnets, precious gems and black sand better than other colors, including black. After only a little practice, a weekend or recreational placer miner using this new pan can work with equal or greater efficiency than the most proficient professional using the old style metal pans or those of black plastic design.

Since Gravity Trap pans can be used for both wet and dry panning, even old stream beds and washes will sometimes produce gold. Built-in riffle traps can be depended upon to trap gold whether water is present or not. True, dry panning is more difficult than wet panning and requires more practice. It can be sometimes more profitable, however, because dry streams that have not seen water for many years – or centuries – can prove to be productive, especially since they were probably passed by during the busier gold rush days. Old timers, remember, with less efficient metal pans, preferred to work with running water because panning there was easier. You may be the first person ever to pan for gold in that specific location. That fact alone can make a trip to the gold fields worthwhile.

If your search is for nuggets in a stream bed or creek containing rocks with high mineral content, a sensitive, ground balancing detector is, of course, essential. In fact, it will probably be impossible to detect nuggets in such a location without this instrument. When you have precisely adjusted the detector's ground balance to eliminate the highly mineralized sands, the instrument will be super-sensitive to any conductive gold nuggets, even the tiniest ones.

Pan Makes Recovery Easier

When engaged in electronic prospecting, a plastic gold pan often facilitates target recovery. Many times it can spell the difference between arduous digging and abandoning a target that has fallen down into a large pile of gravel or large rocks. Whether your target in a rock pile or flowing stream proves to be only a spent bullet or a valuable nugget, the plastic gold pan can make its recovery much faster and easier.

The technique is simple. First, of course, pay close attention to the faintest signals from your detector. They may indicate that the nuggets are small or that they are deeply buried. Second, you must pinpoint each detector response as closely as you can. Then, slip a shovel well under the spot from which the signal came. Be extremely careful now since small objects made of heavy metal always will sink into gravel and may become lost. Place into your plastic gold pan all of the gravel or sand that you scooped up, and scan this sample with your detector.

The value of having a *plastic* pan is now apparent. It would be impossible to scan the sample if the pan were made of metal.

In the plastic pan your conductive target will respond. If there is no response, dump the pan and scan the location again until your detector signals. Pinpoint your target and try to get under it again with your shovel. With a bit of practice, you will be surprised at how quickly you can become proficient in this technique–even when working in three feet of water!

When using these methods to simplify recovery and save time, you will find that use of a pan with the built-in Gravity Trap principle provides not only a quick and efficient tool for sorting through sand and gravel to locate the metallic object that caused your detector to respond, but that also lets you quickly seperate ALL the heavier concentrates from lighter material and facilitates panning that may be necessary.

Your techniques of recovery using a plastic pan are the same when searching a dry wash or placer diggings, except the object will be easier to locate than in a stream. Old dredge tailings may be somewhat more difficult in which to locate targets, and you may lose a few here before you master the technique. Material in which you are working is loose, and an object of heavier metal can easily and quickly work its way deeper into the pile of tailings. Once lost, they are often very difficult or impossible to recover since they simply work their way deeper with each attempt you make to recover them.

You will be tempted to give up and move to a new location. On the other hand, you'll be surprised at how quickly you become proficient and tenacious at such recoveries once you realize that the metallic object you are seeking just might be a beautiful and valuable gold nugget!

Today's electronic detection equipment and improved pans are discovering new gold producing areas daily, and the known producing areas of the past are giving up gold deposits that the old timers overlooked. Fun, excitement and profit of recreational mining are waiting in beautiful gold country. Treasure hunters of today are limited only by desire and time.

The recreational prospector can now achieve excellent results by using a good ground balancing metal detector to locate gold deposits, then panning them with a Gravity Trap gold pan. Whether gold is found in profitable quantities or not, the pleasure of sitting at the edge of mountain stream or in a long-forgotten dry gulch is one that should not be overlooked. Meanwhile, the panner is seeking to produce income with bare hands, knowing full well the chance always exists of hitting "the big one."

Chapter 13

Metal Detector History

The *modern* metal detector really involves a history of only a few years – particularly as it concerns computerized models. All of this is recounted in Chapter 14, Metal Detector Development. The mere detection of metal itself, however, by both mechanical and electrical devices has a long and colorful history that stretches back into antiquity. Very little of this history offers pertinence to the skills or techniques of today's treasure hunter or the modern metal detector. Still, it is part of the story . . . and, an interesting part at that.

Before considering the centuries-old history of metal detecting, it would be well to define some of the feats that a metal detector should NOT be expected to perform as well as to state exactly what it is expected to do.

A metal detector is *not* expected to detect energy emissions from radioactive materials; that is the job of a geiger counter. It is not expected to measure the intensity of magnetic fields; that is done by a magnetometor. It does not point to metal; it does not measure the abundance of metal. A metal detector is expected simply to detect the *presence* of metal and to report this fact.

Thus, from a historical standpoint, a metal detector should be considered simply as an instrument that reports the proximity of metal through some mechanical or electrical means. The importance of such devices along with the variety of their potential uses has long been recognized.

Computerized metal detectors such as Garrett's Grand Master Hunter with its pre-programmed microprocessor controls are as modern as tomorrow's space launch, yet devices used just for detecting metal, in and of themselves, scarcely represent a new scientific development. Now, today's metal detector – the device familiar to treasure hunters – can be described essentially as an electronic apparatus that detects the presence of metal, primarily through the transmission and reception of radio wave signals.

This is the metal detector of *today* and is roughly representative of the basic technology in use for approximately half a century. And, even the first such models represented vast improvement over earlier contrivances rigged up to detect metal. The first such primitive contraption of which historical records exist was probably shaped much like the familiar walk-through detector in use at airports today all over the world.

It made its appearance in the Far East before the time of Christ.

Ancient documents indicate that because a Chinese emperor over two thousand years ago feared assassination, he had developed a metal detecting device using magnets. It was designed to find weapons before they could be brought close to him. How did such a detector operate – you may well ask – centuries before the harnessing of electrical and radio waves in the air? We believe the device was essentially little more than a huge magnet. Legend tells us that it was a crude doorway constructed of a magnetic mineral called magnetite. An iron metal "attractor" was created through heating and/or striking the magnetite with hammers. This jarring and heating caused the molecules to align themselves in the direction of the earth's magnetic field. Woe be to the person who attempted to carry through this doorway such metallic objects as armor, knives, swords or other weapons! The metallic objects would be drawn against the doorway and held fast and the alleged perpetrator exposed . . . and properly dealt with, we can assume!

When President James A. Garfield was shot in Washington in 1881, doctors asked Alexander Graham Bell, inventor of the telephone, to help locate with metal detection equipment the bullet that was lodged in his back and that was eventually to prove fatal. At the time Bell was developing an electric induction device for locating metals. Conflicting reports leave unclear whether Bell ever developed the device and/or whether it was used on the President. They agree on only one aspect of Bell's attempt – it was unsuccessful.

This attempt was described and pictured in an article on Bell in the recent 100th Anniversary issue of *National Geographic.*

The practicality of locating ore bodies in the ground through conductivity, first of their sulfides, is a scientific principle which attracted researchers of varying repute and stature over the years. The idea of finding ore bodies electromagnetically was perhaps first conceived in Nevada in 1904 by Dr. Daniel G. Chilson. In 1909 he turned to known radio transmission/reception techniques, experimenting with short waves.

As industry generally grew in America in the early 20th century, detectors were developed to monitor metal tools and products and control their possible theft by employees and visitors to manufacturing plants. Other uses were also found in industry. A "gate checker" went into service in 1925, designed to help various factories cut down on theft of tools and products. An electromagnetic field was caused to flow across a passageway. Metal carried by persons passing through it altered the electromagnetic field, resulting in a detection signal. The forerunner of today's ubiquitous walk-through detectors, this early model too, could be adjusted to allow small objects to pass through the electromagnetic field without setting off a signal.

At about this same time in the 1920's came the invention of the magnetic balance, a device used for locating underground minerals and metals. From the end of this decade until the coming of World War II, various companies produced metal detector inventions sold as "Amateur Treasure Finders" or "Radio Prospectors." Although little more than gadgets, they certainly represented some refinement of existing theories. The principal contributions of these contraptions came from the imagination inherent in their advertising and marketing.

Throughout development of electronic metal detectors from the earliest days to the present, one factor has been paramount. Certain minerals present in the earth's outer surface plague the effectiveness of such instruments to one extent or another. Not all of the earth's surface minerals are a bother, but the two worst are also two of the most common – iron, which occurs in most of the earth's soil; and salt, which occurs in much of its water. Added to this irony is the fact that salt presents no problem . . . except when it is wet, as in salt water. It then becomes electrically conductive and appears to the metal detector just as if it were conductive metal. Even today, only a few metal detectors can distinguish between wetted earth salt and metal, and only certain models cancel or nullify the troublesome salt.

For the most part, however, the effect of iron minerals can be canceled *(ground balanced)* by a detector without the detector's losing its capability to detect metal. This is accomplished through various methods of circuitry which properly manage the normal electrical phase relationship among resistive, inductive and conductive voltage. Phase shifting is a phenomenon basic to the understanding of electricity. Management of it to "dial out" iron mineralization or other undesirable targets has provided milestones in the development of the modern metal detector. An early milestone led to saving countless lives in World War II with the development of the military mine detector.

During the war rapid advances were achieved in the technology of metal detection as new equipment was developed essentially to locate land mines and similar defensive military weapons. Pressure-sensitive mines buried just below the surface of the earth had become popular weapons.

Facing   Top
The old detectors made fantastic discoveries, even though they lacked the sophisticated circuitry and ground balancing capability of modern detectors.
    Bottom
Bob Marx, right, is one of the many treasure hunters over the world who have relied on Garrett equipment. Here, he and Charles Garrett confer in Florida.
Over
These large rock piles offer excellent territory for a metal detector to find nuggets that were overlooked when the material was dredged from a stream.

In 1945 mine detectors joined countless other items of so-called war surplus available in the public marketplace at a fraction of their cost. Veterans familiar with this equipment were quick to recognize its potential value in locating precious metal ores and buried treasure. Roy Lagal was one of these veterans. The surplus mine detectors were eagerly purchased, mostly by ex-soldiers trained in the use of such equipment who recognized its potential for locating bodies of ore and lost treasure.

And, a hobby was born as well!

A major limiting factor of the World War II mine detectors and those civilian models that went into production after the war to serve the new hobby was the size and weight of the vacuum tubes necessary for their operation. A significant number of tubes was required to provide these detectors with the ability to penetrate ground of any sort. Heavily mineralized areas were, thus, virtually impenetrable.

During the early years after World War II "mine detectors" were used almost entirely to prospect for precious metals. With technological advances, it became evident that metal detectors could find far more than nuggets and ore veins. Soon, they were being used to help locate such other types of treasure as lost coins and jewelry, caches and relics. Metal detectors were used to explore ghost towns and find lost storehouses of treasure.

Facing
Modern detectors have brought a new dimension to coin hunting! The world's first "thinking" detectors, they are finding prizes that were left behind by others.

Over
Treasure hunters are continually amazed by the finds of their new, computerized detectors that hunt deeper and with more sensitivity than earlier models.

Finally, detectors were developed that could be taken deep under the ocean where successful location of lost treasures worth millions and millions of dollars has brought them widspread notoriety.

None of this happened "overnight." In fact, such development, as traced in the following chapter, occurred over almost a quarter century from the 1960's to the 1980's.

By the 1960's, transistors had come into popular use. As transistors replaced vacuum tubes, smaller detectors were possible. Numerous lightweight models of all sizes, shapes and descriptions were offered for sale. Looking back on these early models (Just try to *use* some of them!) is probably humorous to the hobbyist of today. Humorous, too, were the early day automobiles and airplanes. And, imagine this – the first television sets required a huge cabinet of equipment to produce a black-and-white picture not much larger than the palm of a man's hand!

In each of these fields and industries, however, crawling and walking came before running. So, too, in metal detecting.

It was in the 1960's and 70's that real progress began to be made in developing stable and sensitive detectors that featured rudimentary target identification and ground mineral rejection. The Garrett company, founded in 1963 by one of the authors and his wife Eleanor, has been and remains a leader in the development of all types of new and advanced metal detection equipment. This volume's second author has been associated in the development of metal detectors now for some 40 years and has worked with Charles Garrett since the late 60's.

As two who can cite truthfully their service as pioneers in the development of the metal detector, the authors of this book are proud of the record of progess of the metal detector industry, generally, and of that of the Garrett company in particular. In reviewing the early day achievements of the metal detector industry, comparisons should be made with other fields. Just as the Wright Brothers "mechanized kite" flown at Kitty Hawk led to the DC-3 which led to the 747, the SST and the Space Shuttle, so too did the first metal detectors lead to the ADS models which led to the Grand Master Hunter – which the authors truly believe to be the finest metal detector ever manufactured.

Chapter 14

Modern Detector Development

T he story of the modern metal detector is one of continued progress and development over the past two decades. This progress has been especially intense in the past few years as the computerized, microprocessor-controlled instruments were being developed with the pre-programmed Grand Master Hunter standing as a supreme achievement. The most sophisticated metal detector every developed, the Grand Master can be operated with the simple touch of a single button.

As recounted earlier in this book, the early – and, not so early – days of the metal detector industry were filled with what can best be described as gimmicks. They were represented by abbreviations of one sort or another used to describe equipment features. All had meaning, but the meanings of some were more important than those of others.

The Venerable BFO

The letters BFO, which stands for Beat Frequency Oscillator, are important because they are used to describe the first workhorse detector of the modern era. Into the decade of the 1970's the BFO was the detector favored by amateur and professional treasure hunters and, especially, by electronic prospectors.

Veterans still look back on the BFO with fondness. They will tell you that there is something about the instrument that makes you fall in love with it. You respect it for its all-purpose capability, extreme reliability and field ruggedness. Even its audio was especially pleasing! As the detector industry struggled in its growth pains, the BFO stood head and shoulders above the many so-called "modern" detectors that were being introduced to the accompaniment of complicated abbreviations. It became an article of faith among veteran THers that when their new and modern detector failed, they could always depend upon the BFO.

Unfortunately, however, they could not depend upon the BFO for much depth or for cancelling out any ground minerals. Those were its failings – especially the complete impossibility of ground balancing it. Many old-timers can amaze you with their ability to use the BFO in mineralized ground, but the simple fact of the matter is that any of today's quality detectors – with precise ground-balancing capabilities – are far superior to the finest BFO ever made, no matter who was using it.

Parallel with the BFO came the development of the TR (Transmitter-Receiver) detector, also known as the IB (Induction Balance) detector. Although its circuitry was completely different from that of the BFO, it suffered the same complaint concerning mineralization. Some manufacturers tried to alleviate this weakness by adding so-called "ground control" features. These were nothing more than gain controls or sensitivity adjustments. When the effects of mineralized ground were lessened, therefore, sensitivity was also lessened.

A major plus with the TR was its Discriminate mode, which enabled the operator to "tell" the detector which targets to accept and reject. Understanding the proper use of this Discriminate mode is basic to the operation of today's modern metal detectors.

Still a third basic detector, developed along with the BFO and TR, has come to be known as the VLF (Very Low Frequency) detector. Since all of today's detectors operate in the low frequency range denoted by its name, it could be said that the VLF "won" and emerged as the hobby's primary metal detector.

But the story of the VLF is hardly that simple.

VLF Development

Early commercial metal detectors used an operating frequency of approximately 1 kHz (kiloHertz or cycles per second) which was very close to the Army mine detector frequency. But, as it became apparent that higher frequencies produced better results, not only in depth capabilities but in operation of the Discriminate mode, detectors were designed to operate in the frequency range of 5 kHz to 20 kHz.

All electrical/electronic equipment operates at signal frequencies from zero frequency (Direct Current) up to millions of Hertz (cycles)classified into many frequency bands or "groups." For instance, we hear frequency sounds in the 20 Hertz to 20,000 Hertz band. This audio band also includes three smaller classifications. The ELF (Extremely Low Frequency) band includes all frequencies from 30 Hertz to 300 Hertz. Next comes the VF (Voice Frequency Band) from 300 Hertz to 3,000 Hertz (3 kHz). The VLF (Very Low Frequency) band includes all frequencies from 3,000 Hertz (3 kHz) to 30,000 Hertz (30 kHz).

The VLF instruments all offered the feature of ground cancellation or ground balancing which was not possible with the BFO or TR detectors. This made the new detector quite popular, especially for those who sought to work over highly mineralized soil. The first VLF instruments were made with only one mode of operation – what we would now call All Metal. It was quickly learned that the Discriminate mode, so popular on the TR instruments, could also be incorporated into the VLF detectors. To achieve the Discriminate mode the early VLF's operated in a somewhat higher range than today's instruments.

Almost all major manufacturers now, however, produce detectors that operate in a frequency range of approximately 6 kHz. One exception is Garrett's Groundhog curcuit on its Gold Hunter detector which is approximately 14 kHz. This higher frequency is especially suited to finding small nuggets deep in highly mineralized soil.

Ground cancellation describes the ability of a metal detector to ignore iron earth minerals, either automatically or by manipulation of controls. Do not confuse ground elimination of iron earth minerals with the ability of a detector to ignore salt water (wetted salt). For those detectors which permit manual ground balancing, iron earth minerals are eliminated in the All Metal mode; elimination of salt water is added when the Discriminate mode is selected. Discriminating detectors with automatic

153

ground balancing eliminate both iron earth minerals and wetted salt simultaneously.

Discrimination lets the metal detector operator select those metals considered desirable. These choices are "dialed into" the detector through trash elimination controls. The detector then reports to the operator through changes in audio volume or meter deflection concerning which targets are being found. Good (or desirable) objects cause the audio and visual indicators to increase; junk (undesirable) objects cause a decrease.

From ADS to Computer

Garrett's ADS system helped pioneer all of these improvements that had been developed for the metal detector and incorporated them all, including sophisticated target identification meters. Dual discrimination controls added still another dimension to that popular feature. Garrett engineers recognized that the next logial step for metal detector development would incorporate microprocessor control.

Garrett engineers began working in the mid-1970's on the first patent for computerized circuitry based on microprocessor control of a metal detector.

Roughly a decade was to pass, however, before this feature was incorporated into a Garrett detector. The company recognized that such an instrument had to be *easy to use* if it was to be accepted. They realized that a hobbyist wanted to spend time looking for treasure, not constantly programming a detector. **Pre-programming was essential.**

This was finally achieved by the Garrett team in 1987. The patent was granted, and the Grand Master Hunter was born. This detector represented a true breakthrough since it incorporated all the developmental advancements known to date in metal detectors.

Simply stated, the computerized Grand Master Hunter of Garrett or any other company is a *thinking machine*. It performs literally millions of analytical computations – almost simultaneously – that make circuitry adjustments which formerly were made by hand by the hobbyist. As the searchcoil receives data, it is fed into microprocessor circuitry in digital form and compared with the "mind" of the computer, which is a vast storehouse of data that has been stored in the computer at the factory. Thus, knowledge that formerly was required from the operator is now stored in the computer itself. The Grand Master Hunter, therefore, can make adjustments that were once required of the operator.. Not only does it make these adjustments automatically, but they are made instantaneously – when they are needed, not when the need for them is noticed by a hobbyist.

As the detector is scanned, it continually performs self tests; that is, it self-adjusts to achieve optimum operating performance for all conditions, including battery power available, temperature changes and even the possible aging of electronic components that might cause "values" to change. Target and ground mineral data coming through the detector's searchcoil is, first, compared with operator requirements dialed in by the operator, then processed by the instrument's microprocessor-controlled circuitry to produce the proper audio and meter indications. False signals, caused by conventional detector "back reading" are eliminated. Even large, shallow objects are properly read on the meter, with the precise audio tone given.

Garrett engineers have known for years that different ground mineral conditions cause different discriminating performance. The Grand Master Hunter has various scanning methods stored in its memory bank. As earth mineralization changes – even while the detector is being scanned – the detector automatically readjusts itself to use the optimum discrimination method.

What do all these developments – that culminated in the Grand Master Hunter – mean to the treasure hunter? Simply stated, they permit the operator to *improve performance in all phases of treasure hunting.* Greater depths and considerably more discriminating accuracy is possible. Additionally, a detector can now automatically monitor external and ground conditions to maintain all its circuitry at optimum levels. Of course, some of these capabilities can be achieved only with the computerized microprocessor-controlled detectors and are not possible with conventional instruments. It has been said, however, that operator "mistakes" can be virtually eliminated with the "thinking" detectors!

155

Metal detectors underwent vast improvements during the quarter century that began about 1960. Ground balance, discrimination, greater depth, pinpointing and improved stability were but a few of these developments. Over the past two or three years the new, computerized instruments have far overshadowed all earlier developments. *Today's instruments are as superior to those of just a few years ago, as our finest instruments of the 1970's were to the Army mine detectors!*

A new era in treasure hunting has begun. Wonderful prizes are already being discovered that could never have been found before. "Worked-out" areas are producing vast amounts of *new* discoveries. Professional performance and detection accuracy are being achieved easily by beginners at levels that have tantalized professionals for years. We have truly entered the era of *hi tech* metal detector performance. Treasure hunting and electronic prospecting will never be the same again.

Chapter 15

Practical Theory

This chapter will explain in layman's language each of the basic features and components of the modern metal detector. We will speak primarily of the new Garrett Grand Master Hunter microprocessor-controlled computerized detector, the ultimate instrument available today. You will notice, however, that our comments are applicable to most other modern detectors, manual-adjust or automatic.

Some of the comments in this chapter may seem elementary to the veteran treasure hunter. It is important, though, that anyone using a modern metal detector have an accurate and thorough understanding of every component of the instrument and the capabilities it can offer. New metal detectors may look the same as the older models. After all, each of them still has a control housing, a searchcoil and a connecting stem. But, believe us when we tell you that today's modern instruments are different. They have opened new doors to the future for treasure hunters. With a modern detector you can make discoveries you never before thought possible. We urge you to learn all that you can about the practical aspects of these new instruments. Learn how they *will help you find treasure!*

Perhaps you have questions that are not answered in this chapter or anywhere else in this book. First of all, we urge that you read your Owner's Manual carefully. It should answer your questions; if not, go back to the dealer from whom you purchased the detector and get the answers from him. If, for some reason, this is not possible, write to the manufacturer. Explain your questions carefully, and you should receive an immediate answer. If you *still* have questions concerning your detector or any other aspect of metal detecting, write to Ram at the address shown in this book, and we will try to help you.

Now, to the practical aspects of your detector. Some of you who have read earlier books we have written – separately, as well as together – on metal detecting may wonder about the circuit diagrams and mathematical formulas that appear to be "missing" from this book. It is true that various of our books have included them as we sought to "tell it all" about metal detecting. Over the years, however, we've had few hobbyists tell us how much they enjoyed these sections of our books or how much the diagrams and mathematical equations had helped them find treasure. So, you'll find this chapter on Practical Theory just a little more *practical* than its predecessors. We intend to talk here about the practical theory of metal detectors as it concerns treasure hunting.

For the answer to any additional technical questions you might have we recommend Charles Garrett's *Modern Metal Detectors.*

Battery Check

Your detector should offer some sort of visual or audible check to let you know that the batteries powering it are in satisfactory condition. In Garrett instruments this is accomplished by either tones or meter indications each time the detector is turned on.

Read your Owner's Manual to make certain you know the way your detector reports on battery condition, and always pay attention to this report. Never take your batteries for granted! Run-down batteries are the primary cause of "detector failure." Dealers and factory technicians alike are often amazed at the many instruments returned to them for *repairs,* when all that is necessary is a fresh set of batteries.

Audio

Because the audio setting is one of the most important adjustments you make in learning to use a metal detector, we stongly urge your close attention to this subject. We believe that maximum capability can be achieved only by adjusting the audio volume until you hear just the faintest sound coming from your speaker or headphones. This is very important! This faintest sound that you can hear is the detector's most sensitive operating point. It is called your *threshold,* a common term in all metal detector literature. You will notice that when you set the threshold to a very faint sound and then plug in your headphones, the threshold may be too loud. Simply turn the audio control knob slightly to reduce the sound level back to your faint threshold level.

Perhaps many of you prefer to operate a detector with what is called a *silent threshold*; that is, with absolutely no sound coming from the speaker or headphones. If you are determined to use this silent threshold, we urge that you achieve it by setting your audio to a slight level of sound, then backing off just enough to achieve silence. This adjustment insures that any detection sound will go above the silent threshold you are maintaining. Be certain to check occasionally to make certain that you remain at this audio level just below sound.

Discriminate Mode

The Grand Master Hunter and many other of the modern detectors feature two basic modes of operation – Discriminate and All Metal. In the Discriminate mode the operator can utilize discrimination controls to designate the type metal targets that are desired.

Several Garrett detectors, as well as those of other manufacturers, feature dual discrimination controls. They offer multiple selectivity and the ability to reject and accept targets in both the ferrous – iron, of course – and non-ferrous ranges. The two controls split the full range of discrimination between ferrous and non-ferrous. Detection of iron objects such as nails, some foil, iron bottlecaps and small pieces of junk is controlled by one knob. The other control governs discrimination of such non-ferrous items as aluminum pulltabs and aluminum screwtops.

Each of the two controls operates independently of the other. The setting of one has no effect whatsoever on the other. If you wish to detect all ferrous materials, rotate its left control to zero (fully counterclockwise on Garrett detectors). As you advance it back to the right to higher numbers, you will reject more and more ferrous materials. The control operates cumulatively; that is, if you have it set at bottlecap rejection, most nails and some foil will be rejected along with bottlecaps. We urge that you advance this control *no farther clockwise* than necessary to eliminate the troublesome ferrous junk material in the ground you are searching.

Operate your non-ferrous control in the same manner. When it is turned fully to the left, few of the non-ferrous materials will be rejected. To eliminate, say, pulltabs, rotate the control clockwise to the manufacturer's suggested setting for them. Keep in mind, however, that there seem to be as many different kind of pulltabs as there are canning companies. Some few pulltabs, especially those that are bent or broken, seem to be acceptable to any detector at any setting. Set your controls for those you are finding just in the area where you are hunting.

Here's how to set these dual discrimination controls – or a single discrimination control – precisely: Collect examples of that junk you want to reject – a nail, bottlecap, pulltab and, perhaps, small pieces of iron trash. Place your detector on a non-metallic, preferably wooden, surface with the searchcoil at least three feet away from all metal. Make certain you are wearing no rings or jewelry on your hands or arms that could be detected. Rotate both (or only one, if your instrument is so designed) control knobs fully counterclockwise to their lowest settings. Turn the detector on and listen for the beep telling you it is ready to operate. Adjust the audio control for threshold sound.

Pass the iron bottlecap across the bottom of your searchcoil about two inches away from it. Your detector will probably make a signal. Rotate the ferrous control to the approximate bottlecap reject position or the setting suggested by your manufacturer, and pass the cap across the searchcoil's bottom again. You should hear nothing more than, perhaps, a slight blip. You may be able to rotate the control counterclockwise back to a lower number and still not detect the bottlecap. Practice so that you can set your control as far to the left as possible because you always want to use the lowest setting that is required.

Using the same technique, adjust the non-ferrous control just far enough clockwise that you do not detect the standard aluminum pulltab. This should be approximately the manufacturer's suggested setting point. Again, let us stress that you should rotate these controls no higher than necessary to reject the junk items in the ground where you are searching.

Dual discrimination controls such as those on Garrett's Grand Master Hunter and some other detectors offer a greater dynamic adjustment range than single controls. You have more resolution which allows you to set the controls precisely to reject specific junk targets. A most important feature is that you can reject most aluminum pulltabs while accepting the majority of gold and silver rings. When searching for rings on a pulltab-infested area such as a beach, set your non-ferrous control no farther clockwise than necessary to eliminate most of the pull-tabs. Rings with a higher conductivity and, especially, mass than pulltabs will be accepted. Remember, however, that some rings will fall into the lower, or ferrous, range. Thus, dual discrimination lets you select rings that register both "above" and "below" pulltab rejection. So, don't advance either control any farther clockwise than absolutely necessary.

There is another important reason for setting your discrimination controls conservatively. When a modern detector locates a junk target that you have asked it to discriminate against, it cancels out this junk target with a negative audio response that you normally cannot even hear. As you know, however, good targets generate a positive response which you love to hear. If both positive and negative targets are beneath your searchcoil simultaneously, the two reponses tend to cancel one another, and you may miss a good find. Of course, the situation is rarely this simple. Depth of targets, their metallic content, size and many other factors must be considered. So, simply remember this: **never use more discrimination than you absolutely need.**

All Metal Mode

This mode detects all metals with no discrimination. It also detects to the greatest depths possible with most detectors. The All Metal mode is used extensively by cache and relic hunters and by electronic prospectors. Of course, it can be used by coin hunters or anyone who seeks extreme depth and is willing to forego discrimination – except for those detectors which offer metered discrimination.

After activating the All Metal mode in your detector, your

161

next chore will be to ground balance it, a job which the circuitry of most detectors performs automatically in the Discriminate mode. Ground balancing modern detectors, however, is usually a simple task, but is one that is vital to cancel out minerals in the soil.

Check your Owner's Manual for complete instructions, but the directons for ground balancing in the All Metal mode given elsewhere in this chapter will prove applicable for most detectors.

Meters

Meters, of course, vary from detector to detector. The target identification meter on the computerized Grand Master Hunter has five different scales. In discussing it, we will seek to answer any questions you might have concerning the meter on your detector as well as provide some guidance concerning what you should expect from a meter generally. The bottom scale on the Grand Master's meter concerns battery condition. NiCads are reported in the white section, and all others in the yellow range. Batteries are checked each time the detector is turned on or anytime the ON touchpad is pressed. When the meter swings into the yellow range, or white section for NiCads, batteries are satisfactory.

The meter scale above batteries reports coin depth in inches, calibrated for an 8 1/2-inch searchcoil. Other size searchcoils result in slight variations. Also, objects considerably smaller or larger than coins may not be measured accurately.

Next scale above is the identification range. Three classes of metal conductivity – iron, gold and silver – can generally be read very accurately. Keep in mind, however, that there are count-less numbers of alloys that combine these and other metals. Readings on this scale will depend upon which metals are pre-dominant in detected targets.

The next scale upward, target identification, includes such words as foil, bottlecap, nickel, pulltab and penny. Below penny you will see the letters ZN for zinc and CU for copper to differ-entiate between the two kinds of pennies. Most coins will read accurately regardless of how they are lying in the ground. Tar-gets not shown on this scale will be identified according to metal content. Of course, large objects like aluminum cans will read high upon the scale since aluminum is a very good conductor. Whenever the target is quite deep or very small, the meter will either not respond or will move up only slightly into the extreme left range, which is marked "Out of Target ID Range." We suggest that such a target always be dug since it could be a coin buried very deeply.

At the top of the meter on the Grand Master is a linear scale marked zero to 100. This roughly measures the mineral content of the ground over which the detector is scanning, but it also has several other uses. At the top of the meter are the letters OT marking the ore test calibration point for testing gold, silver and other precious metals.

Headphones

We cannot overemphasize the importance of using headphones. Use a set of *high quality* headphones whenever possible, and you will surely reap the benefits. Only when safety or some other overriding considerations demand that you listen to outside noises as well as your detector should you use the instrument's internal speaker.

We emphasize again that you should not underestimate the value of using headphones at all times. Not only will you detect deeper targets when using headphones, but you will be able to ascertain the characteristics of detector signals more readily. Plus, there is an added bonus: using your detector with headphones requires far less battery power; therefore, your batteries last longer.

Detection Depth (Sensitivity)

You will usually want to detect as deeply as possible no matter what detector you are using. Depth of detection, however, will always depend upon many factors that can interfere with performance. Ground mineralization is the most obvious, but it represents only one variable. When you turn on most detectors, they are operating at about 75% of depth capacity. For greater depth, increase the controls marked "Depth" or "Sensitivity." If your detector has a meter, be sure to notice it. As you increase sensitivity (depth) your meter indicator should register theses commands by changing accordingly.

For example, Garrett's Grand Master in the Discriminate mode offers 35 detection depth increments between the 50% depth level and 100%. In the All Metal mode it offers five detection depth increments. Most hobbyists, especially beginners, want to operate at maximum depth *all* of the time. Sometimes, however, conditions just won't permit this . . . just like conditions won't permit you to drive your automobile at maximum speed all the time. Modern detectors permit you to be extremely precise in selecting the maximum level possible

Facing   Top
Precise ground balance and proper discrimination are but two of the modern detector tools used by this treasure hunter to help him make a discovery.

Bottom
Dual ferrous and non-ferrous discrimination controls on modern detectors permit the treasure hunter to be precise in the targets he chooses to reject.

Over   Top
Garrett's Scuba-Mate offers an underwater platform on which the Sea Hunter can be mounted. Only the headphone cord connects it to the diver.

Bottom
An old fort in the Caribbean provided the location for a number of discoveries by Charles Garrett shown searching here with a pistol-grip detector.

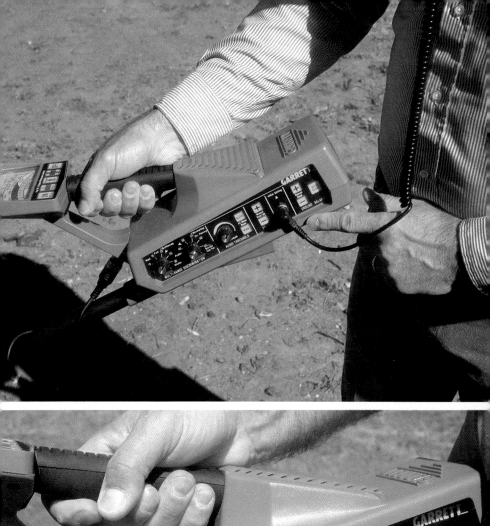
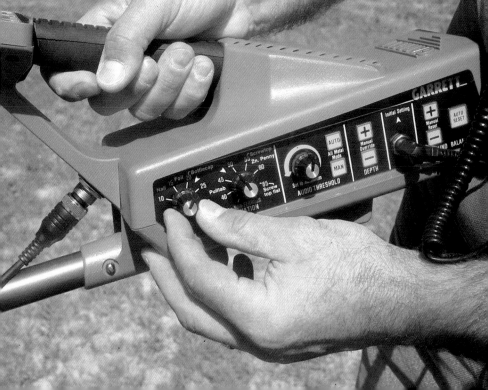

under the electromagnetic and other conditions where you are scanning.

Occasionally, you might be called upon to locate a newly lost object such as a ring or small pin. Most likely this object will still be in the grass or sand at a very shallow depth since it hasn't had time to become covered. Since you do not require maximum depth for such a search, set your detection depth control accordingly. You should hear extremely clear audio signals and recover the lost item quickly.

Ground Balance

Ground balance is probably the most important feature of a metal detector. Many would argue depth, and some might opt

Facing   Top
Headphones that enhance detector signals for treasure hunters are vital to insure that maximum performance is achieved by a modern instrument.

Bottom
A wealth of information is available from the meter of a modern metal detector. Treasure hunters should study to become familiar with these capabilities.

Over
Searching for nuggets in a stream can be a profitable and pleasant pastime for the electronic prospector.

for discrimination. The simple fact remains, however, that without precise ground balance treasure hunting would not be possible in most soils. There would be too much mineralization. Beginning treasure hunters, accustomed to good automatic ground balance on modern detectors, may take this feature for granted.

Please don't! Some of the old-timers are still in awe at the ease of today's ground balancing. They will tell how important it is, and as you progress in the hobby there will be many times when precise ground balance will be demanded if you are to achieve optimum results. Learn how to ground balance your particular detector. It will probably as important as anything you ever do after you turn on the instrument and set the audio.

The following instructions will apply to Garrett detectors and to most other instruments as well, but you are urged to read your Owner's Manual carefully concerning ground balance. Hold the detector with the searchcoil away from any metal and about two or three feet above the ground. If your instrument offers automatic ground balancing, activate the appropriate controls for automatic ground balance. Now, lower the searchcoil to operating height.

Ideally, the automatic ground balance will have proved effective and the audio signal will remain the same or change only slightly, indicating that your detector is properly ground balanced. Test it by scanning the searchcoil over the ground. If it is erratic (audio sounds *not* caused by metal targets), proceed to adjust it manually. Some computerized detectors do not have manual ground balance circuitry. The Grand Master Hunter includes this feature.

The procedure for ground balance, of course, differs from instrument to instrument. Basically, however, if the audio signal grows louder as you lower the searchcoil down to operating height, turn down the ground balance control dial or press the button on your manual ground balance controls marked minus (−) several times. Lift your searchcoil again and lower it to operating height. If the sound level now decreases, you have made too great a negative adjustment. It will be necessary for you to press the touchpad marked plus (+) a few times or turn up your control dial. Remember that with the dial on a Garrett or

170

other quality detector you are dealing with a 10-turn control for precise adjustment. Don't be afraid to turn it several times! Repeat these procedures until the audio does not change or changes only slightly when the searchcoil is lowered to operating height.

When searching extremely mineralized ground, we recommend that you operate the searchcoil two inches or more above the ground. You will not lose depth, but will actually detect deeper because ground mineral influence is greatly reduced.

Pinpointing

Most veteran treasure hunters pride themselves in their ability to pinpoint targets using only the basic controls of the detector. Yet, modern instruments make pinpointing so much easier that we old timers should swallow our pride and take advantage of the electronic assistance available to us. Who knows? The time we save might let us recover that "big one" that's always just around the corner.

Check the Owner's Manual for your detector; a button or trigger on your detector will usually activate the pinpointing circuitry. After you have detected a target, move the searchcoil off to the side, press and *hold* the Pinpoint control and scan your target area again. You will notice that signals have probably grown sharper to aid you in more precise pinpointing.

Some units do not have audio electronic pinpointing but feature a metered pinpointing capability.

Here's a tip for the ultimate in target pinpointing. Once you have determined the surface location where you believe the target to be buried, place your searchcoil lightly on the ground above it and activate the Pinpoint control. Continue holding this control and slide the searchcoil back and forth over the target at the same operating height. You will notice very slight blips when the target is directly beneath the center of your searchcoil. If you can't notice these blips at first, perhaps you have elevated the searchcoil from its level when you first activated your Pinpoint control. Try the procedure again a few times. Maintain constant searchcoil height, and you'll be amazed at your precise electronic pinpointing ability of modern detectors. Warning: this technique requires *practice* and *study* of your manual.

171

Measuring Coin Depth

Once you have pinpointed a target, you can determine the depth of coin-sized objects very accurately, using the meters on many modern detectors. Make one or more scans over your pinpointed target while activating the Pinpoint control. Release the control at any time and observe your meter. It will indicate the depth of your target in inches, if it is a coin. This is a very accurate way to determine target depth and this system cannot be fooled. Generally, it is best when pinpointing to place the searchcoil upon the ground, activate your Pinpoint control and slide the coil back and forth a few times over the target. Then release the touchpad to read the most accurate depth.

Tone Adjustment

This feature permits an operator to adjust the tonal quality produced by a detector's speaker or over its headphones. Some consider this feature important.

Frequency Separation

This feature permits a treasure hunter to operate a detector on each of several different frequencies, which can be selected. This is important if several detectors are operating in close proximity. If they are all operating on the *same* frequency, they will "talk" to each other when they get close together.

Chapter 16

Detector
Configurations

O n most metal detectors the control housing is mounted on one end of a stem with a searchcoil at the other. The cable connecting the searchcoil and control housing is wound around the stem, and the detector is carried by a handle either at the end of the stem or on the control housing. Many variations of this configuration can be seen, but some detectors are still as bulky and unwieldy as they were 30 years ago. They are ugly to look at and difficult to handle.

On the other hand, Garrett and other progressive manufacturers have devised exciting new designs for their modern metal detectors. Let's examine some of these configurations:

Standard

As noted above, the basic (or standard) configuration of most detectors features a control housing attached to the handle and stem with a cable wound around the stem to the searchcoil. This configuration is both historic and traditional and can be traced back to the earliest commercial models.

On such modern detectors as the Garrett Master Hunter series, comfort and ease of handling are of far more concern than custom or tradition. These detectors feature what is often called a "wrist action" model. You should always *consider balance and weight of utmost importance* in the selection of a standard configuration detector. Newcomers to the hobby and veterans alike will often want to spend long hours in the field with their detectors. An instrument that is heavy or improperly balanced can leave muscles sore and dispositions ruffled. On the other hand, lightweight models can be used for long periods with little mental or physical fatigue.

Balance, too, is important in lessening strain on hand and arm muscles. We define balance as the ease with which a detector rests in the hand when held in the normal operating position with searchcoil extended. Little effort should be required to hold the searchcoil in the air at operating height. Light weight and good balance will result in minimal fatigue experienced both during and after treasure hunting.

The Grand Master Hunter, for example, is particularly well balanced. Just a single finger placed beneath the forward end of the handle permits the detector to float perfectly in the air at the proper scan angle. The operator is not forced to raise the coil continually to keep it balanced. The detector's handle permits complete grip comfort. In fact, it is not necessary to **grip** the handle. Simply, let the detector float comfortably as it is cradled in the curled fingers of either hand. The meter is placed in a protected position on the end of the handle for both visibility and rugged performance. Touchpads, the major controls of this particular instrument, are placed right at the hand of the operator where they can be reached easily and operated instantly by the touch of a thumbtip. The speaker is placed for maximum sound direction to the ears of the operator while being protected from an environmental standpoint. Below the rubber handle grip, the speaker is truly hidden. The cross-louvered area is actually an opening for a multiple slit beneath which is mounted a waterproof speaker.

Pistol-Grip

As manufacturers sought to make their detectors easier to handle, the pistol-grip model was developed. This type of detector usually features a built-in extension arm rest as well as the easily grasped handle that furnishes its name. Excellent balance and light weight are both generally features of this type instrument. These enable a hobbyist to count on scanning for hours without tiring. The pistol-grip stem is designed both for hunting on land as well as in shallow water.

This type of detector can be moved about easily as the operator carries out scanning techniques. With the detector as an "extension" of the arm and hand, its searchcoil stem lies along the same line as the forearm. Motion is accomplished without thinking since operation is almost as simple as "pointing a finger."

The popular Garrett Freedom series features this design. Easy access controls are on the front of the detector facing the operator, and those models with electronic pinpointing place the trigger control at the "trigger finger" of the operator.

Pistol-grip configurations are popular for searching on the beach and in the surf. They are light and easy to handle, especially with the control housing either mounted on the body or on some sort of float. With these detectors to be used in or near the water, it is important that *adequate environmental protection* be provided for the instrument's housing. On a non-submersible model the housing must be either mounted on some sort of float or high on the body. Of course, Garrett's AT4 which features an arm rest and pistol grip on the searchcoil stem is hip mounted, but it is designed with the necessary environmental protection to be submersible in shallow water.

Hip-Mount

This configuration features the control housing on a belt around the waist or slung over the shoulder with the searchcoil on an adjustable-length stem. An armrest is also usually supplied. This configuration to which some standard detectors can be converted is designed to relieve the arm of weight and to protect the control housing in some surf-hunting models. This model is ideal for those hobbyists who desire, for one reason or another, to minimize the weight that is carried by the hands and arms.

On any model that is to be worn on the hip, it is important that all controls be easily accessible in this configuration.

The Garrett AT4 Beach Hunter is manufactured as a hip-mounted detector to provide added comfort and greater flexibility in its use in shallow surf and on the beach. Beach hunting can be a hot and sometimes difficult job, and working in the surf always offers the challenge of dealing with water currents. Placing the bulk of the control housing on an operator's hip not only removes this burden from the hand and arm but makes the detector far less unwieldy for easy operation in moving water or shifting sands.

The use of the hip-mount configuration requires searchcoil cable that is considerably longer than that required for the wrist-action or pistol-grip configurations. Ninety inches is the normal length. If you purchase a detector and plan to use it mounted on your hip, make certain that the searchcoils you buy along with it have the longer cable.

Using searchcoils with standard-length cable on a hip-mounted detector will require an extension cable two to three feet in length with mating conductors. Fitting out your detector in such a way offers no particular problem except in waterproofing. Make certain that your connectors are protected if they will be submerged or exposed to dampness.

Chest-Mount

These models, while quite functional, have never enjoyed popularity. Usually, the control housing is suspended with a cross-shoulder strap that holds the housing flat against the upper chest. This configuration is generally designed to protect the housing while hunting in shallow water.

When converting your detector to chest-mount configuration, make certain that all controls are easily accessible.

Flotation-Mounted

This configuration locates the control housing on a flotation device when the detector is used for hunting in calm water such as lakes, ponds or gentle surfs. The number and variety of such devices are limited only by the imagination of water hunters.

Underwater Designs

The Garrett Sea Hunter detector is manufactured for use to depths of 200 feet. Designed for efficient land, surf and under-water hunting, it is built in the hip-mount configuration but the control housing can also be mounted on an arm, leg or the upper chest. Of primary concern to many underwater hunters, how-ever, is keeping such bulk from the body. Divers in "tight" situations such as shipwrecks do not want to be encumbered with large objects strapped to the body.

Garrett produces its "Scuba-Mate" for such divers. This accessory is a mobile underwater platform on which the detector is mounted. Only the headphone cord connects it to the diver. In any emergency, therefore, the rig can be scuttled easily. It can also be lowered to a diver, when necessary, so that swimming movements are not encumbered with this added weight.

Similarly, many underwater hunters find it awkward to have a long stem attached to the searchcoil. Digging and retrieving discoveries is difficult when the searchcoil is several feet in front of you. Swimming to the point of detection is inconvenient, as is thoroughly searching the hole from which a target is removed. With a short (approximately 12″) stem, a diver has the searchcoil at arm's length and can merely reach out to dig a target or fan the sand away. Checking the hole is also easy.

Pocket Scanners

One of the most effective searching tools in the treasure hunter's bag of tricks can be the pocket scanner or so-called "hand-held" detector, still another configuration in which metal detecting equipment is designed and manufactured. Of course, all treasure hunting metal detectors are held in the hand, but the pocket scanners can literally fit into an individual's palm.

The detectors obviously can penetrate only about 18 inches, but they are very handy for treasure hunters trying to search difficult places – particularly inside a house. You can quickly scan ceilings, walls and floors with such a detector. They are also useful for checking in deep holes that you have dug to make certain that you don't leave a second treasure after you remove the one you have found. They are also good for pinpointing all targets within the hole itself. The little hand-held instrument can reach places where even the smallest searchcoil will not go.

When the instrument is calibrated for ore sampling like Garrett's Little Treasure Hunter, electronic prospectors find a pocket scanner quite handy for checking ore samples. Once again, this little detector can scan in tight places where no searchcoil could be used.

An effective pocket scanner should be a tool used by all modern treasure hunters.

Chapter 17
Searchcoils

Searchcoils are an integral – and vitally important – component of any modern metal detector. We urge that you never take your searchcoil for granted or overlook its importance. The finest detector in the world can be no better than its searchcoil.

It is obvious that no detector can operate without a searchcoil. What is not so obvious is that the type of searchcoil used with a modern detector will determine how effectively it will perform the task that has been assigned to it . . . indeed, how effectively it can perform *any* task. You wouldn't use cheap or retread tires on a high-speed motorcar. Don't expect an intricate modern detector to operate effectively with a low quality searchcoil.

Just what is a searchcoil? Let's continue our motorcar analogy and define it in automobile terms. Searchcoils basically have the same function as wheels on a car. Wheels take power from the motor and interface between the automobile and the ground. They roll along, take bumps and shocks to permit the car to perform its function of getting to a destination. Searchcoils take power from the control housing via the searchcoil cable. They are the interface between the metal detector and the ground. They take bumps and shocks as they scan to permit the detector to perform its function of finding targets.

Most searchcoils operate with electromagnetic transmitter and receiver antennas embedded within the searchcoil. The searchcoil is mounted on the lower end of a stem to be scanned over bare or water-covered ground, rocks or a specific object. An invisible electromagnetic field generated by the transmitter winding flows out into the medium that is present – soil, rock,

179

water, sand, wood, air or whatever. Further description of this process is given in Chapter 4.

Because design and construction of searchcoils is so critical, manufacturers of modern metal detectors are quite concerned with it. Searchcoils are made of wire and plastic which can contract and expand with normal temperature changes thus affecting the electrical parameters of searchcoils. Instability, drift and erratic operation will result unless extreme care has been exercised in the design and manufacture of each searchcoil.

An important aspect of any searchcoil is the effective scanning width it provides through the mounting of its antennas and other design criteria. Just because the shell of a searchcoil is eight inches in diameter does not mean that that coil will scan down into a circle of this size. We urge you to experiment. You may be amazed at the limitation of some detector searchcoils.

Also very important to a modern detector is the searchcoil's ability to pinpoint accurately. Modern detectors all feature pinpointing, with circuitry especially designed for such a purpose. Without the proper searchcoil, however, this special pinpointing circuitry is useless.

Searchcoils come in many shapes and sizes. Roughly speaking, the smaller the searchcoil, the smaller the object that can be detected. Larger searchcoils are primarily designed to detect deeper targets, but they can locate tiny objects at any depth as well.

Environmental Protection

To be effective any searchcoil should offer some degree of environmental protection, primarily against rain and other forms of water. Even if a hobbyist doesn't intend to hunt in shallow water, searchcoils should be able to resist moisture that will occasionally be encountered. It might be well to review here the "waterproofing" designations in general use with searchcoils. *Splashproof* indicates that operation will not be affected if a small amount of water gets on the searchcoil, such as moisture from wet grass. *Waterproof* means that the searchcoil can be operated in heavy rain with no danger from the moisture. *Submersible* indicates that a searchcoil can be submerged as deep as the cable connector without affecting the detector's operation.

Make certain you understand the water-resistant capabilities of your searchcoil. This is important! All Garrett searchcoils are fully submersible. That means that they can not only "resist moisture," but operate effectively when submerged in water to the connector and sustain no water damage, no matter what detector is used. Of course, the AT4 is designed for total submersion and the Sea Hunter can operate in water depths of 200 feet.

Submersibility of *any* kind may not be feasible for the searchcoil you are using. Make certain of its environmental integrity before you risk ruining it by dunking it in the water!

All searchcoils are not alike. They vary widely both in quality and in what they will enable a detector to find. Simply stated, a good searchcoil is vital to the success of a metal detector. No hobbyist should seek a "bargain" in purchasing a searchcoil!

Electronic Shielding

Searchcoils must also be electronically shielded to protect them from electrostatic interference. The most common type of such interference will probably come from wet grass. Garrett and most other manufacturers shield the windings of modern searchcoils with what is called a Faraday Shield. Once again, however, we urge you not to take this protection for granted. Question before you purchase a searchcoil, or test it for yourself in this manner. With the audio threshold of your detector properly adjusted, drag a handful of wet weeds across the bottom of its searchcoil. If very noticeable changes occur in the sound, the searchcoil does not have effective shielding. Slight audio changes may be expected when wet grass passes over the top of a searchcoil.

Rugged Treatment

Your searchcoil must be light, but it must also be sturdy and capable of rugged treatment. You want a tough exterior that will not abrade or tear on rough ground.

Searchcoils must withstand the greatest abuse of any detector component because they are constantly being slid across the ground, bumped into rocks and trees, submerged in water and generally mistreated in every way. Consequently, manufacturers must exercise great care in construction. We urge that hobbyists exercise similar care through the use of searchcoil covers, known sometimes as skid plates.

Garrett's new Crossfire searchcoil models feature a special innovation designed to permit them to take rugged treatment and keep providing top performance. This is a "tri-point" suspension that isolates the transmitter/receiver windings from shock and encapsulation stress. The winding arrays literally "float" within the searchcoil and do not respond to every bump and shock. The resulting stability permits winding the antennas for peak performance. The Crossfire coils also offer 100% scan width because of the effective method in which antennas are mounted. Effective pinpointing is guaranteed for targets at any depth.

All Garrett searchcoils have been subjected to rigorous testing in the laboratory and out in the fields and waters where hobbyists will expect them to perform. Field testing by the authors of this book and other treasure hunting experts has been a key ingredient in the successful performance of all of Garrett's equipment from the company's earliest days.

Again, let us urge that you never underestimate the importance of a searchcoil on your detector. Remember, also, that searchcoils are manufactured in various sizes for specific reasons. Investigate and experiment! Learn which searchcoils operate best on your instrument for each specific type of treasure hunting.

Even though the term is used rather loosely (even by us, we must admit!) there is truly no such thing as a **standard** searchcoil. The term "standard" is an absolute misnomer because the coil that operates ideally in a park may be next-to-useless on a junk-filled beach. The searchcoil that finds coins would rarely be used to search for a cache. Searchcoils come in a wide range of sizes.

Let's review the various sizes and types of searchcoils that are available for the modern metal detector.

7-9″ Diameter

This size searchcoil is furnished with most detectors which is proper because such a searchcoil is the best size for general purpose use. These searchcoils are usually lightweight, have good scanning width and are sensitive to a wide range of targets. Small objects can be detected, and good ground coverage can be obtained. Shallow scanning width is approximately equal to the diameter of the searchcoil. Depth of detection is satisfactory for most targets with a searchcoil of this size.

3-5" and Smaller

At Garrett we refer to this size searchcoil as our Super Sniper. Its intense electromagnetic field gives good detection of small objects, and its narrow pattern permits excellent target isolation and precise pinpointing. Depth of detection is not as great as that of larger sizes. Remember, however, that a searchcoil illuminates everything in the search matrix. In high junk areas it is possible to find targets with a Super Sniper that would be masked by junk signals if a larger coil were used.

12" and Larger

Searchcoils of this size, while able to detect coin-size objects at great depths are also classified as the smallest searchcoils to be used for cache and relic hunting. Precise pinpointing is obviously more difficult with the larger sizes, and their increased weight usually necessitates the use of an arm rest or hip-mounted control housing, especially when the detector is used for long periods.

Of course, you will want to use this larger searchcoil when searching for deep caches or relics. But, when you are coin or nugget hunting, how do you know when to switch from your general purpose size to the larger searchcoil? Suppose you locate a target in the fringe area of detection. You know from the weak audio signals that you are at the outer edges of your detector's capability with this searchcoil. By using a larger size, you will detect deeper.

Elliptical

These searchcoils whose name indicates their shape are no different from most other searchcoils even though they are not round. Such searchcoils are designed primarily for electronic prospecting where they can reach between rocks into places that could not be searched effectively by a circular searchcoil.

Depth Multiplier

Special two-box attachments are manufactured for extra deep searching. The Depth Multiplier, which Garrett calls the "Bloodhound," multiplies the depth that an instrument can detect. For example, if a large cannon or safe can be detected to a depth of seven feet with a 12-inch searchcoil, the Bloodhound can locate it to perhaps twice that depth.

The Depth Multiplier is recommended when searching for money caches, large relics, safes, cannons, ore veins and mineral structures. The Depth Multiplier takes no more battery power than smaller searchcoils, but the added depth capability is one that will benefit any treasure hunter.

Some of these attachments are known by the term "TR" because of their special transmitter and receiver windings. These require special adjustment when installed on a detector. On the other hand, Garrett's Bloodhound is easy to use. It attaches in a matter of seconds to the control housing of Garrett's computerized Grand Master Hunter or any of the Master Hunter series. No tuning or ground balancing adjustments are required. Simply turn the detector on in its All Metal mode with manual tuning. Hold the Depth Multiplier about one foot off the ground as you search. Since this special searchcoil seeks out only large objects, you'll not be bothered by small pieces of junk or ground mineralization.

Searchcoils literally "connect" a metal detector with the ground beneath it. No modern instrument can operate effectively without a good one!

Facing
Hip-mounted detectors such as this Beach Hunter take weight away from the arm and hand and permit long hours of searching with less muscle fatigue.
Over
With the pistol-grip Garrett Freedom detector serving as a virtual extension of the THer's arm, moving it around is a simple as pointing a finger.

Chapter 18
Clothing and Accessories

E ven with the most sophisticated modern metal detector, locating a target is only a part of the treasure hunting story. First, you must have the equipment to get you *to* the treasure for which you are searching. Then, you must have the neccesary equipment that will enable you to recover it.

This chapter will consider many of the various accessories that are *necessary* in modern treasure hunting. Note the word, necessary. There is no end to the number and kind of accessories available to treasure hunters today. Your metal detector dealer . . . sporting goods shops . . . hardware stores . . . all have implements and items of clothing galore that might possibly help you be a more effective treasure hunter . . . and a better looking one too! In this chapter, however, we will consider only those accessories that we consider vital to the "arsenal" of the modern treasure hunter. And, we will make a few observations about clothing that have proved helpful to us over the years. A

Facing
The standard, wrist-action Grand Master Hunter almost **floats** in a treasure hunter's hands with controls at the touch of a thumbtip.
Over   Top
The Super Sniper coil searching this fireplace goes where other coils could not and can scan near metal support plates to locate hidden caches.

        Bottom
Searchcoils, such as these Crossfire models from Garrett, are vital to the operation of any metal detector. Stable operation under rugged conditions is a must.

good rule of thumb is to put your money in a good, modern detector, not in frills and accessories.

Headphones

Always use headphones!

There can be no ifs, ands or buts about this accessory. You should use headphones whenever you are searching with a metal detector. Unless there is some sort of powerful mitigating circumstance, such as necessary safety precautions, we urge that you always use headphones. Of course, this is even more important when ambient noise is especially loud, such as on the beach, near traffic or in fast-moving water.

Stated in simple terms, headphones will enhance your audio perception by bringing sound directly into your ears while masking outside noise interference. You can hear weaker sounds and detect deeper targets when quality headphones are used. Prove it to yourself by burying a coin and scanning for it, using only your detector's speaker. Then, put on headphones and scan the same spot. Believe us; you'll be amazed at the difference. Better than just believing us is trying it yourself!

Headphones come in a variety of sizes, shapes and configurations, the most popular type being the stereo ear muffs. Many detectors do not have volume controls, but headphones with volume controls allow a wide degree of audio volume adjustment while not degrading the actual signals of the detector itself.

Detector designers all know that reducing sound volume of a detector's signals results in loss of detection depth and sensitivity. Reducing circuit gain limits the sharp, quick audio turn-on vital to proper discovery. Even on those detectors with volume control, manufacturers will usually recommend that volume be set to maximum. Most detectors are operated at full volume with the audio tuning adjusted so that only a faint threshold sound is coming from the speaker. Only when a target is encountered does the sound rise quickly from threshold to maximum volume. Headphones permit thresholds to be set even lower, further improving performance.

A mono/stereo switch on headphones is desirable. When both earpieces are in use, mono operation lets the sound come into both. When use of only one is desired – perhaps to leave the other ear free for those mitigating safety sounds – only the one ear piece will be operating if you are switched for stereo operation. If you want this feature, be sure to check the headphones before you buy them; some will send sound into both earpieces regardless.

You can find headphone sets of varying sizes and weights. Those that are large and cover the ears with acoustical material eliminate more outside sound and let you hear detector signals more clearly. They are also heavier and can sometimes become hot and uncomfortable. You might wish to consider purchase of one of the lightweight sets for use in hot or humid areas, or simply when the heavier headphones become bothersome.

Any set of *quality* headphones is better than trying to determine detector signals from the speaker alone.

The key word in that preceding paragrah is quality. Headphones will last you a long time. Invest in one or more good pairs. You'll want right-angle plugs because they lessen the possibility of breaking one of them off in your detector's headphone jack. This happens more than you imagine with plugs that stick straight out. Since plugs come in different sizes, you'll want to make certain you have the right size for your detector. You'll also want a coiled card that will keep itself out of your way when you are scanning.

One final note on headphones. When you are using them, people around do not know what you are hearing. Perhaps this may seldom be important to you; on the other hand, you might not always know just what those people standing around are doing or what they are listening for!

Batteries

Whenever you find your detector inoperable . . . whenever it will not respond properly . . . whenever it is just plain "broke" . . . *check your batteries*. Without a doubt, weak or worn-out batteries are the single, major cause of detector failure. We don't intend to suggest that new batteries will always cure your detector's problems, but batteries are usually one source of the problem whenever a detector stops working. Always carry a fresh set of batteries with you whenever you are hunting. Even though your detector's meter or tone indicator reports that your batteries are satisfactory, test the detector with fresh batteries any time it fails.

Whenever you change batteries, make certain the new ones are inserted properly and test satisfactorily. Regularly check your detector's battery and power system. Inspect terminals for corrosion and tightness. Look for wires that might have been pinched when they were pulled out earlier for changing batteries. These pinched wires can set up a problem that leads to failure later in the field. If your detector's battery compartment (or any part of the instrument, for that matter) is sealed at the factory, *never try to open it.* You may ruin your detector; at best, you'll void your warranty.

Types of Batteries

Carbon zinc batteries cost the least and deliver current for the shortest period of service. They operate most efficiently at fahrenheit temperatures from about 32 (freezing) to just over 100 degrees. They are more prone to leak than any other type of battery.

Heavy duty (zinc chloride) batteries are generally more expensive than carbon zinc, but will give longer service. They are more prone to leak than alkaline or NiCad batteries.

Alkaline (alkaline manganese) batteries cost more than carbon zinc and heavy duty types, but they generally provide current for considerably longer periods of time. Furthermore, they last longer in storage, are less susceptible to leakage and perform better in extreme temperatures. Their use is probably cheaper in the long run than carbon zinc or heavy duty types.

NiCad (nickel cadmium) rechargeable batteries are more expensive than the other types because manufacturers claim that they can be recharged one thousand times. Maximum life and best performance can be achieved if they are used often and recharged immediately at room temperature.

Always remember that NiCad batteries will generally "take a set" if repeatedly used the same length of time. For example, if you repeatedly use your NiCad batteries for only one hour before recharging them, they will "take a set" of one hour, and that is the maximum length of time they will ever deliver current. That is why it is often good to let your NiCad batteries run down completely before recharging. At least once every three months completely discharge and restore a full charge to NiCad batteries to extend their life.

NiCad batteries will generally power a circuit only 40 to 50% as long as carbon zinc batteries. For example, if your detector will operate for 20 hours on carbon zinc batteries. NiCads will power it from eight to ten hours. Since NiCad operating voltage is less than that of the other types, they register at a lower level on meters and lights designed to check batteries on detectors.

Digging Tools

In one respect, treasure hunting is much like every other field of endeavor. If you are wise in selecting the correct tools and use them properly, you will magnify both your results and the pleasure you gain from achieving them. In other words, take the right tools into the field with you and know how to use them properly. Now, the *right* tools may vary widely, depending on what you're searching for and where you're looking for it. What you'll need to carry with you onto a beach of loose sand is far different from what you'll need for prospecting in a mountain stream.

For basic treasure hunting we suggest four basic pieces of digging equipment:

– A heavy duty garden trowel;
– A small shovel;
– A light-weight pick with a flat blade on one end.
– A screwdriver-like pointed probe

Use the trowel and probe first until you become proficient with them. Eventually, you'll be able to use the pick to dig up your target with a single whack without even kneeling down sometimes. Of course, the shovel will come in handy for any deep or hard digging.

Pinpointing is essential before you begin using any digging tool. Know what size object you will be searching for and how deep you can expect to find it. Before you can graduate from the probe and trowel to a pick, you'll have to perfect your pinpointing.

Of course, to pinpoint properly you'll may need to use a longer probe . . . some sort of long, slender and sturdy rod. Metal detector shops sell various kinds of probes, or you can make your own. When searching for caches, you'll need a long probe. When you make such a probe, a small "bullet" welded on the tip will reduce its amount of drag going in and out of the ground. Only slightly larger in diameter than the probe rod itself, the bullet should be pointed to facilitate insertion into the ground. You'll need a solidly attached handle to remove a long probe.

Be especially careful when using a probe, particularly if you might locate a coin. You do not want to scratch the coin with your probe. That's a reason for using one with a dull point rather than one with a sharp point or a knife.

The screwdriver can be used alone to recover shallow targets, and it is ideal for digging through tree roots. Many coin hunters swear by a screwdriver as a digging tool. They maintain that they can dig up coins with less damage to grass and sod by using a screwdriver than with any other digging tool.

Sometimes you'll have to cut and remove a plug of grass in a well-manicured lawn to get to your target. There are probably as many techniques here as there are veteran coin hunters, but they all agree (most of them, anyway) on several things:

– Cut your plug deep with plenty of earth;
– Try to "hinge" the plug back rather than removing it completely;
– Leave your coin below the plug itself so that you can dig it out of the soil beneath rather than damage the plug;
– Make certain the plug is replaced solidly so that it won't be pulled out by a lawnmower;
– Try to damage surface grass and roots as little as possible.

Before leaving the subject of digging, we must remind all treasure hunters again to *fill your holes*. Simple courtesy and safety are but two reasons for leaving every location in which you search in *better* condition than when you found it. Unfilled holes give our hobby a black eye and present a menace to anyone walking by. Later in this book we discuss the Code of Ethics for treasure hunters with metal detectors. Making certain that you always fill your holes is one of the most important parts of it.

Pouches

You'll need something to carry your discoveries in . . . and also to carry out the trash you find. *Never* put trash back into the ground. Who knows? You might search the area again yourself some day!

A waist apron with at least two or more pockets provides the simplest means of carrying out discoveries and other items. Some treasure hunters use three-pocket pouches – one for trash, one for keepers and one for more valuable discoveries. These apron pockets or pouches need not be covered unless you'll be moving around actively, or you permit the pockets to get too full.

Pouches should be waterproof to prevent soiling your clothes, and the fabric should be sturdy for rigorous use and withstanding lots of weight. Most pouch styles can be mounted on a belt. Many treasure hunters use a web-type military belt. On it they hang not only treasure/trash pouches, but a canteen, first aid kit, shovel and other digging tools, camera and various other items.

Beach and surf hunting require some specialized recovery tools in addition to those discussed here. For more complete information on them we recommend Charles Garrett's *Treasure Recovery from Sand and Sea.*

Similarly, prospecting has its own requirements, and we recommend the new book we have just published, *Modern Electronic Prospecting,* and Roy Lagal's new book, *Weekend Prospecting.*

Cases

We strongly recommend that you purchase a carrying case for your detector(s) at the earliest opportunity. No matter if you carry it only in the trunk or back seat of your automobile, a case will protect your instrument and its controls. Two types are available, a hard cover carrying case or a soft vinyl bag. Either is satisfactory, depending upon your preference and requirements.

Modern metal detectors are rugged instruments. They are built for outdoor use in varying types of weather, depending on environmental protection. Never forget, however, that a detector is a scientific, electronic instrument. Always handle it carefully and protect it as much as possible from mist, rain or blowing sand. Always keep your instrument as clean as possible. Try not to put it away dirty, and never fail to rinse any salt water off it immediately after using.

Clothing

Just a few words about clothing for the modern treasure hunter are appropriate for this chapter on accessories. Your main goals in selecting clothing are to *be comfortable* and to *protect yourself* against the elements. In other words, stay warm in the winter and cool in the summer and dry at all times!

Remember that you'll be in the outdoors for long hours at a time. Wear enough heavy clothes in the winter to protect against cold weather. In dry climates this cold can "sneak up" on you before you know it.

In hot weather don't forget to wear a hat and to take along plenty of water so that you won't get dehydrated. Don't neglect protecting your neck and arms against the hot rays of the sun. Lotions that screen you against the sun have been improved to such an extent that just a single application will remain effective for hours, regardless of how much you perspire. And, they're not oily or sticky any more!

Gloves are an important item of any treasure hunter's wardrobe. You'll need them for protection, especially when digging. Never dig barehanded, even in loose dirt or sand. There are simply too many pieces of broken glass and old razor blades lurking around that can inflict severe injuries quite easily. Any individual who works with his or her hands knows the importance of gloves. You'll be using your hands when you hunt for treasure with a metal detector; don't neglect them.

This brief summary of accessory equipment is by no means meant to be all encompassing. Many of you probably have some piece of equipment that is not mentioned above, yet that is considered vitally important to you. So be it! Use whatever equipment is familiar to you, and that you consider important.

Our list seeks to point out just those accessories that we believe are *necessary* to any treasure hunter, novice or veteran.

Chapter 19

Points to Ponder

I n this chapter we'll consider four more aspects of this delightful avocation of treasure hunting:

1. The universal nature of the pastime and the fact that it can more than pay for itself;

2. Health aspects of this recreation;

3. The Metal Detector Code of Ethics;

4. Some of the legal implications of the hobby, including a simple Search and Salvage Agreement.

Enjoyable, Profitable, Universal

We believe that there is no more enjoyable – or profitable – way for men and women to spend their leisure hours than hunting for treasure with a metal detector. The hobby . . . the thrill of discovering treasure . . . appeals to people of all ages. It is an utterly universal outdoor pastime! Treasure can be hunted with equal intensity anywhere on the face of the earth or under its waters. There are absolutely no geographic or topographic boundaries or barriers.

All limitations are completely self-imposed. Each individual decides how much energy is required to excel at the hobby, and the decision can be changed from day to day or even from one minute to the next. The hobbyist can hunt for hours a day or for just a short while; the hunting may be strenuous or involve little exertion. Hidden wealth can be sought with equal vigor and satisfaction at exotic foreign locations or literally in one's own backyard.

Hunting for treasure with a metal detector is an ideal hobby for young people . . . full of energy and curiosity . . . with a desire for adventure and excitement. It can lead them to exciting new locales and introduce them to new people and places. The hobby is equally – if not even more – suitable for mature men and women . . . yes, *senior citizens,* whose health permits (or requires) light outdoor exercise and who have maintained their zest for adventure. Treasure hunting offers the opportunity to satisfy this quest for excitement and mystery without lengthy travel or elaborate equipment.

Thus, it's an ideal hobby for the entire family, where several generations can enjoy the outdoors and each other together. Each can work and play at his or her own pace without interfering with anyone else. Rewards are immediate and obvious. Keeping score is unnecessary, but the winners are obvious!

Truly, just the seeking of lost treasure proves so fascinating that everyone is a winner. The hobby also offers financial rewards as well as the benefits of healthy exercise and outdoor activity. Finally, nothing can compare with the sheer thrill of discovery – whether it be that first coin . . . a ring . . . a gold nugget . . . a lost outlaw cache. The joy and excitement enrich both the spirit and the pocketbook.

Pays for Itself

To say that treasure hunting is a hobby that can truly pay for itself tells only part of the story of this great pastime. Certainly, the hobby can repay the cost of its necessary equipment . . . many times over, in fact. Even though the price of a quality modern metal detector might seem somewhat high to a novice, the rewards that come with it can be great indeed. These include not only those rewards of the spirit but actual bounty that can bring immediate financial benefit. When you find a modern coin with no numismatic value, what could be better than spending it at once!

But, there is more to the hobby of treasure hunting than financial rewards. Unless you intend to become a professional treasure hunter, we urge that you not undertake this hobby only for the value of those objects you can find. You'll be missing much of the enjoyment of treasure hunting if you only pursue it for financial gain – and you might be disappointed!

Oh, if you learn your lessons from this book properly, use the right equipment and practice diligently, you'll find plenty of what you are looking for . . . be it coins, jewelry, gold nuggets or whatever else. We've seen too many become successful to doubt that you, too, can find treasure. But real success comes only with the *joy of the hunt*. Dr. J. Garland Starrett is a fictional character created by Charles Garrett to tell about some exciting real-life treasure hunts in which Charles has participated. The first novel, *The Secret of John Murrell's Vault*, has just been published. At the close of this novel, Starrett makes a statement about the hunt just completed that is so pertinent that every treasure hunter should heed it:

"What a delightful treasure hunt it was! But, aren't they all? If nobody gets hurt or spends money he or she can't afford, every treasure hunt is a genuine pleasure."

We couldn't have said it better ourselves!

Health Aspects

People of all ages and in all stages of physical condition enjoy the hobby of hunting with a metal detector. They roam parks, ghost towns, beaches and gold mining areas in search of treasure. Some occasionally complain of muscle pains or aching joints after they have used metal detectors for long periods of time. It's been our experience that most such complaints come from newcomers to the hobby who have become captivated by it. They've found their first coin or ring and can't get enough of metal detecting. As a result, they swing the detector 10 or 12 hours those first days. The following mornings they naturally wake up with a good case of cramped muscles and maybe a sore joint or two. After a short time the soreness disappears, and off they go again.

In some of our earlier books we have offered the following suggestions for lessening the danger of strained muscles:

– Select the **proper equipment**, including accessories, considering weight and balance in relation to your physique and stamina level;

– Strengthen hand, arm, back and shoulder muscles through a moderate **exercise** program;

– Before beginning each day's detecting activity, spend a few minutes **warming up** with simple, easy stretching exercises;

– During the actual activity of metal detecting, use common sense in how you employ your muscles; use **correct scanning techniques** and take an occasional break.

Let's expand on proper scanning techniques. First, keep a firm footing at all times; never try to scan while standing on one foot or in some other unnatural position. Make all movements as natural as possible. If you find yourself scanning on steep hills, in gullies and other unlevel places, keep good balance, take shorter swings and don't let yourself get in an awkward position.

Grasp the metal detector handle lightly. If you've selected an instrument such as the Grand Master with light weight and excellent balance, you'll scarcely know you're holding a detector. Slight wrist and arm movements in scanning are necessary when you are moving the searchcoil from side to side in short swings. If you swing the detector widely, use a method that is natural and one that causes a minimum amount of unnecessary wrist movement. Let your entire arm "swing" with the detector.

Occasionally, change hands and use the other arm to swing the detector. If you ever feel yourself tightening up, stop and rest. Most likely, however, you'll get enough rest just digging up the targets you find. Thus, getting a little rest becomes just another reason to find more treasure!

Oftentimes, the discovery of another coin or piece of jewelry is better medicine than any salve or liniment!

Most important of all, use common sense and take care of yourself! *There are no "time limits" to metal detecting. You have the rest of your life.*

Code of Ethics

Filling holes is but one requirement of a dedicated metal detector hobbyist. A sincere request that we make to every user of a Garrett detector is to leave each place searched in better condition that it was found. Thousands of individuals and organization have adopted a formal **Metal Detector Operators Code of Ethics:**

"– I will respect private and public property, all historical and archaeological sites and will do no metal detecting on these lands without proper permission.

"– I will keep informed on and obey all laws, regulations and rules governing federal, state and local public lands.

"– I will aid law enforcement officials whenever possible.

"– I will cause no willful damage to property of any kind, including fences, signs and buildings, and will always fill holes I dig.

"– I will not destroy property, buildings or the remains of ghost towns and other deserted structures.

"– I will not leave litter or uncovered items lying around. I will carry all trash and dug targets with me when I leave each search area.

"– I will observe the Golden Rule, using good outdoor manners and conducting myself at all times in a manner that will add to the stature and public image of all people engaged in the field of metal detection."

Policing this code is an important job of the scores of local metal detector clubs organized over the nation. Clubs varying in size from a few members to hundreds meet regularly for fellowship, to share adventures and to compare their success in the field and water. At the same time, these sincere hobbyists seek knowledge of new developments in the science of metal detecting and try to remain abreast of the rapidly changing laws and regulations that govern their hobby.

If you can't find a club locally, write to Garrett headquarters and let us help you find one. Or, better yet, help you organize one!

Some Legal Aspects

This section is by no means intended to offer legal advice. In fact, the best advice we can give anyone with a legal problem is, *"See a lawyer."* Here, instead, we want to raise just a few legal points that you should consider before going out into the field with your modern metal detector:

Local regulations – Make certain that you are aware of them, especially in parks and similarly regulated vacation and recreation spots. We've heard of equipment actually being confiscated when it was used illegally. And, remember, that ignorance of the law is no excuse! Ask those in authority, and when special permission is required, get it in writing.

Trespassing–Don't. It's that simple. Always heed "no trespassing" signs and never knowingly invade the property of others without getting permission. In some areas the very fact that you are trespassing voids your claim to any part of treasure that you might discover, regardless of other laws governing its ownership.

Finder's Keepers – There may be some truth in this old statement, but there are certainly exceptions. Finder's Keepers may not be appropriate for an object you discover on private or posted property if the landowner decides to dispute your claim. On the other hand, Finder's Keepers generally applies to any owner-not-identified item you find when you are not trespassing, when you are hunting legally on any public land and when the rightful owner cannot be identified. Of course, anyone can claim ownership of anything you find; it may then be left to the courts to decide the rightful owner.

Treasure trove – In the United States this is broadly defined as any gold or silver in coin, plate or bullion and paper currency that has been found concealed in the earth or in a house belonging to another person, even when found hidden in movable property belonging to others such as a book, bureau, safe or a piece of machinery. To be classed as treasure trove the item(s) must have been lost long enough to indicate that the original owner is dead or unknown. All found property can generally be separated into five legal categories:

Abandoned property, as a general rule, is a tangible asset that has been discarded or abandonded willfully and intentionally by its original owners. Thus, it becomes the property of the first person who discovers and desires it. An example would be a household item such as an appliance discarded into a trash receptacle. If the trash collector (or anyone else, for that matter) decides to take the appliance, they can do so legally.

Concealed property is tangible property hidden by its owners to prevent observation, inventory, acquisition or possession by other parties. In most cases, when the property is found, courts order its return to the original owner. Sometimes the finder is given a small reward, more for his honesty in reporting the find than for the effort of discovery.

Lost property is defined as that which the owner has inadvertently and unintentionally lost yet to which he legally retains title. Still, there is a presumption of abandonment until the owner appears and claims such property, providing that the finder has taken steps to notify the owner of its discovery. Such a case might arise when someone finds a lost wallet that contains documents identifying the owner. It is the general rule that such property must be returned to its owner, who pays a reward if he so desires In fact, in almost every jurisdiction a criminal statute exists that makes it a crime to withhold "lost" property.

Misplaced property has been intentionally hidden or laid away by its owner who planned to retrieve it at a later date but forgot about the property or where it was hidden. When found, such property is generally treated the same as concealed property with attempts necessary to find its owner. When this is not possible, ownership of the property usually reverts to the occupant or owner of the premises on which it was found with the finder being awarded some amount of the object's value.

Things embedded in the soil generally constitute property other than treasure trove, such as antique bottles or artifacts of historical value. The finder acquires no rights to the object, and possession of such objects belongs to the landowner unless declared otherwise by a court of law. Generally, courts divide the value of the find between the property owner and the finder.

Finally, don't forget income taxes – federal, state and local. Monetary gain from any treasure found must be declared as income in the year that it is received; that is, when you SELL the items, not when you FIND them. Applicable expenses can be charged against the gain from any treasure discovered. Simply stated, however, tax laws require declaration of all income from treasure hunting.

Written Agreements

Avoid disappointment and heartbreak! Execute a written agreement with property owners and even with your hunting partners when your treasure is possibly a large one or if your fellow hunter(s) is new to the hobby. It has often been our experience that so-called verbal agreements aren't worth the paper they're written on, especially when valuable treasure finally appears.

Speaking of property owners, make sure that your agreement is reached with ALL of them. We've known of cases when a treasure was claimed by a wife after her husband had given permission to search the property. Also, you should be certain that your "owners" are not just *renting*. When dealing with governmental agencies and the stakes are high, consider using an experienced attorney.

Following is an example of a simple agreement between property owners and a treasure hunter. Depending on the property laws applicable to a site or the magnitude of the treasure sought, a more detailed, custom-drawn agreement might be necessary.

Facing
High atop a Colorado mountain peak, Charles Garrett puts the Grand Master Hunter through its paces, comparing it with competitive models.
Over
The latest instructional tools from detector manufacturers are training videos that illustrate the "how to" of modern treasure hunting techniques.

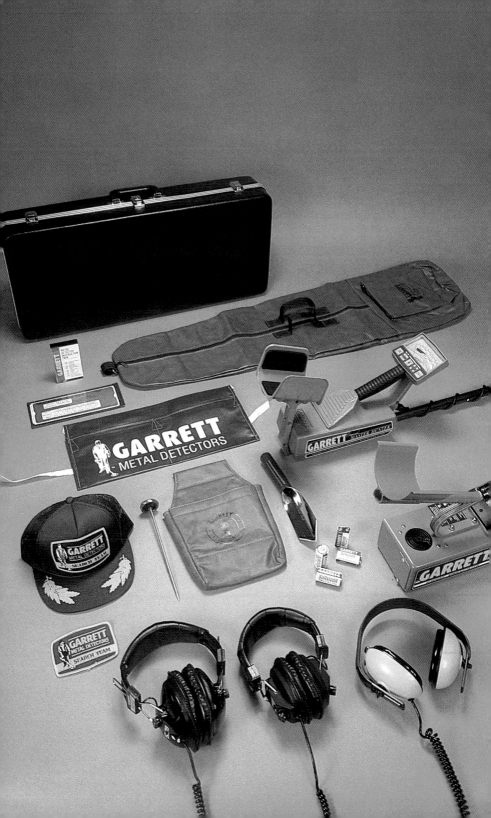

Facing   Top
A Garrett detector found this medallion in England.
Believed to have been worn by King Richard III in the
15th century, it sold for two million U.S. dollars.

Bottom
Proper digging tools are vital for recovering coins
and other discoveries, especially when they are
located beneath trim and neatly manicured lawns.

Over
Detector manufacturers and others offer a wide
range of accessories. Each treasure hunter must
decide which are imporant to his or her type of
hunting.

Search and Salvage Agreement

This agreement, dated the _____ day of _____, 19 ___, between _____, hereinafter known as the Property Owner(s), and _____, hereinafter known as the Salvagor. In consideration of the Salvagor's undertaking to devote his time and equipment in a search of the premises described as

the Property Owner(s) hereby agree that the Salvagor shall receive as compensation for his services _____ (designated percentage) of all money, jewelry, artifacts and

discovered by the Salvagor, subject to the laws of this community. The Salvagor is given full authority to work in, on or about, said premises at any reasonable time, subject only to such notice as the Property Owner(s) may require in advance of such work. Each party waives any possible claim against the other for liability for any careless or negligent act or omission of the other arising out of or in the consequence of the search herein provided for. This agreement shall be effective for a period of _____ from the date thereof.
Executed at:

_____   _____

Salvagor   Property Owner

                                        _____

                                        Property Owner

Chapter 20
Conclusion

E ncompassing two lifetimes of treasure hunting in a single volume is a challenging task. There is so much that we would like to tell you about this wonderful hobby that we have enjoyed for so many years . . . a hobby that continues to excite and arouse us as it inspires the imagination while invigorating the body.

It is a magnificent hobby. And, you should always remember that it is a hobby . . . that you are participating in it not for profit but for the thrill of the hunt. True, monetary rewards may often exceed your wildest imagination, but don't work for this alone. Accept the financial rewards as an added benefit.

Let yourself delight to the mental and physical challenges of the search. You will find few opportunities today where you are continually confronted with challenges as exciting as this.

Some other final thoughts on our experiences with the use of modern metal detectors for treasure hunting:

–Study your detector and this book; become familiar with both. Let this volume be your guide to the possibilities of the modern instruments. Of course, you must also become intimately familiar with the Owner's Manual for your detector. If you find it unsatisfactory or incomplete, tell your dealer of your dissatisfaction. Work with him until you thoroughly understand all of the *electronic* capabilities of your instrument. This guidebook and your own imagination will let you expand on these electronic capabilities to broaden your treasure hunting horizons to a truly unbelievable extent.

–Experiment with your equipment; carry out a continual program of laboratory and field testing. Now, this need not be as elaborate as it sounds. A "laboratory" can be a table in your

garage or workroom; the "field" can be your backyard. But, we emphasize that you should never accept anyone else's ideas about metal detecting on their face value alone – no matter where the idea comes from (even this book!). Test these new ideas you encounter with your own detector to prove their validity. Try out each technique and procedure to make certain that they work for you and that you can accomplish them with your detector. It would be impossible for us to place enough stress on the importance of *testing ideas for yourself*. Be certain they make sense to you and that they prove worthwhile with your detector before you set out to search for treasure. Great expectations and no experience combine themselves too easily to produce disappointment. Don't be afraid of new ideas. Try them out, first; then, add them to your personal bag of tricks if they prove successful. Let experience be your teacher, and work through experiments and testing to build up this reservoir of experience.

– Be *diligent*: few treasures were ever found easily. One axiom of treasure hunting you should learn to accept at the outset: *to find treasure you must also dig trash*. The longer you hunt with a metal detector, the more vividly you will realize – and understand – the truth of this statement. Diligence must also be applied in your study of treasure hunting equipment and techniques. Never look for "short cuts" or the easy way out. Hard work today will reward you many times in the future.

– Be *persistent*: don't quit. The next scan of your searchcoil may bring to your ears that sound of treasure for which you have long been waiting. Make no mistake about it. Lost treasure abounds! There is a vast amount of it to be found today, and it is being lost even more rapidly than you and your fellow treasure hunters can find it. You may consider this statement difficult to accept. We ask that you just consider the vast number of people in the world and the great amount of valuable possessions that are continually being lost, misplaced, forgotten or generally consigned to oblivion for one reason or another. First, you must locate this treasure that is being lost, through intelligent and diligent research. Then, you can recover it by using the techniques of modern metal detection that you have proven sucessful with your instrument. Be thorough and persistent; these techniques never go out of style!

We have written books and articles, conducted workshops and seminars as we explained to two generations the equipment and techniques necessary to find lost treasure. Untold thousands have followed our advice in fields and mountains . . . streams and oceans . . . all over the world. Many have been successful beyond their most extravagant dreams. All have been given a magnificent opportunity for pleasure and joy. May each of you who reads this book be among this happy throng of treasure hunters when we . . .

See you in the field!

Charles Garrett

Roy Lagal

BOOK ORDER FORM

Please send the following books

- ☐ **Modern Treasure Hunting** **$12.95**
- ☐ **Treasure Recovery from Sand and Sea** **$12.95**
- ☐ **Modern Electronic Prospecting** . . . . **$ 9.95**
- ☐ **Buried Treasure of the United States** . . . . **$ 9.95**
- ☐ **Weekend Prospecting** . . . . **$ 3.95**
- ☐ **Successful Coin Hunting** . . . . **$ 8.95**
- ☐ **Modern Metal Detectors** . . . . **$ 9.95**
- ☐ **Treasure Hunter's Manual #6** . . . . **$ 9.95**
- ☐ **Treasure Hunting Pays Off!** . . . . **$ 4.95**
- ☐ **Gold Panning Is Easy** . . . . **$ 6.95**
- ☐ **Treasure Hunter's Manual #7** . . . . **$ 9.95**
- ☐ **The Secret of John Murrell's Vault** . . . . **$ 2.95**

Garrett Guides to Treasure

- ☐ **Find An Ounce of Gold a Day** . . . . **$ 1.00**
- ☐ **Metal Detectors Can Help You Find Wealth** . . . . **$ 1.00**
- ☐ **Find Wealth on the Beach** . . . . **$ 1.00**
- ☐ **Metal Detectors Can Help You Find Coins** . . . **$ 1.00**
- ☐ **Find Wealth in the Surf** . . . . **$ 1.00**
- ☐ **Find More Treasure With the Right Detector** **$ 1.00**

Ram Publishing Company
P. O. Drawer 38649
Dallas, TX 75238

Please add 50¢ for each book ordered (to a
maximum of $2) for handling charges.

Total for Items $_____

Texas Residents
Add 8% Tax $_____

Handling Charge $_____

Total of Above $_____

MY CHECK OR
MONEY ORDER IS
ENCLOSED, $_____

I prefer to order through
☐ VISA   or ☐ MasterCard   (Check one.)

_____
Card Number

_____
Expiration Date

_____
Phone #

_____
Signature (Order must be signed.)

_____
NAME

_____
ADDRESS

_____
CITY, STATE, ZIP

Ram Publishing Company
P.O. Drawer 38649
Dallas, TX 75238